NATIONAL ACADEMIES *Sciences Engineering Medicine*

NATIONAL ACADEMIES PRESS
Washington, DC

Improving the Intelligence Community's Leveraging of the Full Science and Technology Ecosystem

Committee on Leveraging the Future Research and Development Ecosystem for the Intelligence Community

Intelligence Community Studies Board

Division on Engineering and Physical Sciences

Committee on Science, Engineering, Medicine, and Public Policy

Policy and Global Affairs

Consensus Study Report

NATIONAL ACADEMIES PRESS 500 Fifth Street, NW Washington, DC 20001

This study was supported by Contract No. 10004810 with the Office of the Director of National Intelligence. Any opinions, findings, conclusions, or recommendations expressed in this publication do not necessarily reflect the views of any agency or organization that provided support for the project.

International Standard Book Number-13: 978-0-309-68785-0
International Standard Book Number-10: 0-309-68785-3
Digital Object Identifier: https://doi.org/10.17226/26544

Copies of this publication are available from

Intelligence Community Studies Board
National Academies of Sciences, Engineering, and Medicine
500 Fifth Street, NW, Room 938
Washington, DC 20001

This publication is available from the National Academies Press, 500 Fifth Street, NW, Keck 360, Washington, DC 20001; (800) 624-6242 or (202) 334-3313; http://www.nap.edu.

Copyright 2022 by the National Academy of Sciences. National Academies of Sciences, Engineering, and Medicine and National Academies Press and the graphical logos for each are all trademarks of the National Academy of Sciences. All rights reserved.

Printed in the United States of America.

Suggested Citation: National Academies of Sciences, Engineering, and Medicine. 2022. *Improving the Intelligence Community's Leveraging of the Full Science and Technology Ecosystem*. Washington, DC: The National Academies Press. https://doi.org/10.17226/26544.

The **National Academy of Sciences** was established in 1863 by an Act of Congress, signed by President Lincoln, as a private, nongovernmental institution to advise the nation on issues related to science and technology. Members are elected by their peers for outstanding contributions to research. Dr. Marcia McNutt is president.

The **National Academy of Engineering** was established in 1964 under the charter of the National Academy of Sciences to bring the practices of engineering to advising the nation. Members are elected by their peers for extraordinary contributions to engineering. Dr. John L. Anderson is president.

The **National Academy of Medicine** (formerly the Institute of Medicine) was established in 1970 under the charter of the National Academy of Sciences to advise the nation on medical and health issues. Members are elected by their peers for distinguished contributions to medicine and health. Dr. Victor J. Dzau is president.

The three Academies work together as the **National Academies of Sciences, Engineering, and Medicine** to provide independent, objective analysis and advice to the nation and conduct other activities to solve complex problems and inform public policy decisions. The National Academies also encourage education and research, recognize outstanding contributions to knowledge, and increase public understanding in matters of science, engineering, and medicine.

Learn more about the National Academies of Sciences, Engineering, and Medicine at **www.nationalacademies.org**.

Consensus Study Reports published by the National Academies of Sciences, Engineering, and Medicine document the evidence-based consensus on the study's statement of task by an authoring committee of experts. Reports typically include findings, conclusions, and recommendations based on information gathered by the committee and the committee's deliberations. Each report has been subjected to a rigorous and independent peer-review process and it represents the position of the National Academies on the statement of task.

Proceedings published by the National Academies of Sciences, Engineering, and Medicine chronicle the presentations and discussions at a workshop, symposium, or other event convened by the National Academies. The statements and opinions contained in proceedings are those of the participants and are not endorsed by other participants, the planning committee, or the National Academies.

Rapid Expert Consultations published by the National Academies of Sciences, Engineering, and Medicine are authored by subject-matter experts on narrowly focused topics that can be supported by a body of evidence. The discussions contained in rapid expert consultations are considered those of the authors and do not contain policy recommendations. Rapid expert consultations are reviewed by the institution before release.

For information about other products and activities of the National Academies, please visit www.nationalacademies.org/about/whatwedo.

COMMITTEE ON LEVERAGING THE FUTURE RESEARCH AND DEVELOPMENT ECOSYSTEM FOR THE INTELLIGENCE COMMUNITY

FREDERICK R. CHANG (NAE), Southern Methodist University, *Chair*
MICHAEL A. MARLETTA (NAS/NAM), University of California, Berkeley, *Vice Chair*
LILIAN ALESSA, University of Idaho
TOMAS DÍAZ DE LA RUBIA, University of Oklahoma
VISHVA M. DIXIT (NAS/NAM), Genentech
DONALD DUNCAN, Johns Hopkins University Applied Physics Laboratory
GERALD L. EPSTEIN, National Defense University
KATHLEEN FISHER,[1] Tufts University
JAMES R. GOSLER, Johns Hopkins University Applied Physics Laboratory
LAURA M. HAAS (NAE), University of Massachusetts Amherst
ROBERT F. HALE, Center for Strategic and International Studies
DANIEL E. HASTINGS (NAE), Massachusetts Institute of Technology
FRANCES S. LIGLER (NAE), North Carolina State University
BERNARD S. MEYERSON (NAE), IBM Corporation
LISA J. PORTER,[2] LogiQ
PETER SCHIFFER, Yale University
ANTHONY J. VINCI, Center for a New American Security
MICHAEL S. WITHERELL (NAS), Lawrence Berkeley National Laboratory

Staff

DIONNA ALI, Associate Program Officer
ANITA EISENSTADT, Program Officer
SHENAE BRADLEY, Administrative Assistant (until November 2021)
ANTHONY FAINBERG, Senior Program Officer
CHRIS JONES, Senior Financial Business Partner
MICHAEL NILES, Senior Program Officer (until August 2021)
NIA JOHNSON, Program Officer
MARGUERITE SCHNEIDER, Administrative Coordinator
ALAN H. SHAW, Director (until March 2022)

---

[1] Resigned from the committee October 2021.
[2] Resigned from the committee on May 16, 2022.

INTELLIGENCE COMMUNITY STUDIES BOARD

MARK LOWENTHAL, Intelligence & Security Academy, LLC, *Co-Chair*
MICHAEL A. MARLETTA, University of California, Berkeley, *Co-Chair*
JOEL BRENNER, Massachusetts Institute of Technology
ROBERT CARDILLO, The Cardillo Group, LLC
FREDERICK R. CHANG (NAE), Southern Methodist University
DEAN CHENG, The Heritage Foundation
ROBERT C. DYNES (NAS), University of California (president emeritus)
ROBERT A. FEIN, Harvard Medical School
HUBAN A. GOWADIA, Lawrence Livermore National Laboratory
MARGARET A. HAMBURG, Nuclear Threat Initiative
MIRIAM E. JOHN, Independent Consultant
ANITA K. JONES (NAE), University of Virginia (professor emerita)
STEVEN E. KOONIN (NAS), Center for Urban Science and Progress
CARMEN L. MIDDLETON, The Walt Disney Company
ARTHUR L. MONEY, Department of Defense
WILLIAM C. OSTENDORFF, United States Naval Academy
DAVID A. RELMAN (NAM), Stanford University
ELIZABETH RINDSKOPF PARKER, State Bar of California (Retired)
SAMUEL S. VISNER, The MITRE Corporation
DAVID A. WHELAN (NAE), Cubic

Staff

DIONNA ALI, Associate Program Officer
SHENAE BRADLEY, Administrative Assistant (until November 2021)
BRYAN BUNNELL, Research Associate
JOSEPH CZIKA, Senior Program Officer (until June 2022)
MELVIN EULAU, Senior Program Officer (until March 2022)
ANTHONY FAINBERG, Senior Program Officer
CARYN A. LESLIE, Acting Board Director
LIZA HAMILTON, Program Officer (until April 2022)
NIA JOHNSON. Program Officer
CHRIS JONES, Senior Financial Business Partner
MARGUERITE SCHNEIDER, Administrative Coordinator
ALAN H. SHAW, Director (until March 2022)

Acknowledgment of Reviewers

This Consensus Study Report was reviewed in draft form by individuals chosen for their diverse perspectives and technical expertise. The purpose of this independent review is to provide candid and critical comments that will assist the National Academies of Sciences, Engineering, and Medicine in making each published report as sound as possible and to ensure that it meets the institutional standards for quality, objectivity, evidence, and responsiveness to the study charge. The review comments and draft manuscript remain confidential to protect the integrity of the deliberative process.

We thank the following individuals for their review of this report:

NORMAN R. AUGUSTINE (NAS/NAE), Lockheed Martin Corporation
ANDREW BROWN, JR. (NAE), Diamond Consulting & Delphi Automotive
JOHN C. GANNON, Georgetown University
RUSH D. HOLT, American Association for the Advancement of Science
J. MICHAEL McQUADE, Carnegie Mellon University
DAWN C. MEYERRIECKS, Office of the Director of National Intelligence
WILLIAM H. PRESS (NAS), The University of Texas at Austin
J. PAUL ROBINSON, Purdue University

Although the reviewers listed above provided many constructive comments and suggestions, they were not asked to endorse the conclusions or recommendations of this report nor did they see the final draft before its release. The review of this report was overseen by GRANT H. STOKES (NAE), Massachusetts Institute of Technology Lincoln Laboratory, and SALLIE A. KELLER (NAE), University of Virginia. They were responsible for making certain that an independent examination of this report was carried out in accordance with the standards of the National Academies and that all review comments were carefully considered. Responsibility for the final content rests entirely with the authoring committee and the National Academies.

Contents

PREFACE		xi
SUMMARY		1
1	**INTRODUCTION**	9
	Brief History, 9	
	The Changing Nature of Global S&T, 10	
	What Does It Mean for the IC to "Leverage" the S&T Landscape?, 12	
	Charge for This Study, 14	
	How the Study Was Conducted, 15	
2	**A VISION FOR STRENGTHENING THE IC'S ABILITY TO LEVERAGE S&T**	16
	Existing IC Leveraging of S&T, 16	
	Aspects That Limit IC Capabilities for Leveraging S&T, 19	
	Foundational Steps for Strengthening IC S&T Capabilities, 20	
3	**LEVERAGING THE S&T ACTIVITIES OF OTHER FEDERAL AGENCIES**	27
	Mechanisms for Coordination Across Federal Agency S&T Programs, 28	
	Performers of Federally Funded Research and Development, 29	
	Strengthened Interactions with the National Laboratory System, 30	
4	**LEVERAGING EXPERTISE FROM THE FULL U.S. S&T ECOSYSTEM**	34
	Dynamic Nature of Today's S&T Ecosystem, 35	
	Steps for Leveraging Domestic S&T, 36	
	IC Interactions with S&T Industry, 42	

5 LEVERAGING THE GLOBAL S&T COMMUNITY — 44
Existing U.S. Government Engagement in International S&T Cooperation, 46
Existing Cooperative Agreements with Selected Allies, 48
Existing IC S&T Cooperation with Five Eyes and Allies, 49
Ways for the IC to Enhance Its Access to, and Awareness of, International S&T, 50
Increased Use of International Open-Source Information, 51

APPENDIXES

A Leveraging the Future Research and Development Ecosystem for the Intelligence Community by the U.S. Research Community: Proceedings of a Workshop—in Brief — 57

B Leveraging the Future Research and Development Ecosystem for the Intelligence Community—Understanding the International Aspect of the Landscape: Proceedings of a Workshop—in Brief — 69

C Acronyms and Abbreviations — 81

D Committee Member Biographical Information — 83

Preface

Initial discussions for the study that led to this report took place before anybody had heard of SARS-CoV-2. What the committee planned to cover, and how, was quite different from what could actually be accomplished. By the time the project started, COVID-19 restrictions were in full force. While COVID-19 restrictions affected all National Academies of Sciences, Engineering, and Medicine studies, this study—because it touches on topics and context that are inherently sensitive, even when not classified—was particularly constrained. And that limited the depth of the study's investigations and deliberations compared to what the committee would have wished and what the Intelligence Community (IC) needs.

On the surface, having an unclassified study take place virtually might not seem particularly consequential, but in fact, it had a profound effect on the study and on the conclusions that the study could draw. COVID-19 restrictions fundamentally impacted the depth of the study, and the conclusions and recommendations we were able to make as a committee. The restrictions affected the depth of presenter interactions, the depth of committee interactions, and the depth of insight and nuance that these interactions would have provided. It is for these reasons that this report is more generic than we had intended. Because an in-depth study on important parts of the science and technology (S&T) landscape had been released in 2021 by the Center for Strategic and International Studies,[1] the committee therefore approached its task in part as a supplement to the detailed examination in that document. So this report does not replicate the deep and detailed analysis of that report, but instead expands on it by considering the full range of S&T and going beyond that report's major focus on the IC itself to delve into domestic S&T in industry and academia, and international S&T across all sectors, including allied governments.

Frederick R. Chang, *Chair*
Committee on Leveraging the Future Research and
Development Ecosystem for the Intelligence Community

---

[1] CSIS Technology and Intelligence Task Force, 2021, *Maintaining the Intelligence Edge: Reimagining and Reinventing Intelligence Through Innovation*, Washington, DC: Center for Strategic and International Studies, https://www.csis.org/analysis/maintaining-intelligence-edge-reimagining-and-reinventing-intelligence-through-innovation.

Summary

The agencies within the U.S. Intelligence Community (IC) depend on advanced technology to achieve their goals. The IC's missions require it to have the broadest possible insights into, and access to, scientific research, technology discovery and development, and engineering wherever in the world that may occur. Science and technology (S&T) is a major driver of strategic competition with China, Russia, and other adversaries, commands high priority within the IC, and deserves more attention across all IC agencies.

This observation is not new. A particularly relevant examination appears in the January 2021 Center for Strategic and International Studies (CSIS) report *Maintaining the Intelligence Edge: Reimagining and Reinventing Intelligence Through Innovation*.[1] That report highlighted the need for the IC to relentlessly pursue competitive advantage relative to our principal adversaries, particularly by reimagining and reinventing methods to leverage technological innovation. Its focus was on the use of technological innovations that are enabled by "artificial intelligence (AI) and associated emerging technologies, including cloud computing, advanced sensors, and big data analytics." It pointed to the need for both culture change and the mastery of global S&T advances. It offers more than 100 recommendations for how the IC should respond.

Because most of the recommendations in the CSIS report are directed toward internal adjustments to IC institutions, the current report focuses more on improving external connections, as covered in Chapters 3 to 5—how the IC can better leverage S&T knowledge across the broader government, domestic, and global S&T environments. In addition, this report looks more broadly at the full range of S&T to recommend how the IC can (1) innovate and leverage advances from all fields of S&T; (2) be better positioned to identify, track, and employ S&T advances in the service of the core intelligence missions of collection, analysis, and distribution; and (3) improve coordination of S&T intelligence (S&TI) to advance the core intelligence missions of preventing strategic surprise and causing strategic surprise to adversaries. The need for such a broadened approach also is not new. The "whole of government approach" is generally considered essential to meet serious and growing challenges of foreign origin, not only via direct interference but also in the areas of economic and scientific competition. Similar observations have been

---

[1] Center for Strategic and International Studies (CSIS), 2021, *Maintaining the Intelligence Edge: Reimagining and Reinventing Intelligence Through Innovation*, a report of the CSIS Technology and Intelligence Task Force, Washington, DC, January.

enunciated by diverse experts, for example from the Congress, the private sector, and prominent congressionally directed commissions, including the recent Solarium and AI commissions."[2]

Several external factors over recent decades have heightened the importance of S&T to the IC. The growth of commercial enterprises based on cutting-edge S&T—for example, AI-based software and the use of big data, advanced computing, microelectronics, biotechnology, novel materials, and other developments—means that technology and its applications are progressing rapidly on multiple fronts. And many of these developments are occurring not only in open environments such as universities and government laboratories but also in commercial firms. At the same time, this increased commercialization of S&T is also becoming more international. That is especially true for fields such as life sciences research and development (R&D), telecom, and computing-based technologies that have taken root in multiple countries. On top of these trends is China's ascension as an economic and political rival, and as a global leader in S&T. The IC in general needs a more effective mechanism than it currently has to gain this knowledge and understanding—what the committee calls "S&T awareness"—and to feed relevant information where it is needed within the IC for S&TI and for mission support. In order to assimilate a full picture of significant global scientific activities, especially those deriving from developments in emerging technologies, one cannot—and should not—rely on intelligence agency sources alone. A useful understanding of current, vital progress in many fields must be filled out by continual input from other U.S. agencies, laboratories, federally funded research and development centers (FFRDCs), research universities, and the private sector. Unlike the IC, these all have ready access to expert scientists, who interact with foreign partners and with national and international scientific bodies and networks.

These trends combine to imply a pressing need for the IC both to enhance its awareness of these advances and to innovatively incorporate these capabilities into the various IT missions. Those capabilities must support the IC in several ways:

- Generating and maintaining S&T awareness—which includes S&T from all sources that could be useful—to enable recognizing and tapping external S&T advances to support IC operations, including collection, analysis, support, and management. These external advances are in addition to the IC's internal S&T investments.
- Carrying out S&TI—which consists of collecting and analyzing information about current S&T advances, particularly by adversaries or that may be used by adversaries—in order to prevent surprise, to be able to surprise adversaries, to negate adversarial advantage, or to provide advantage to the United States. Whereas information for S&T awareness can be provided by a broad range of technically trained people, collecting and analyzing for S&TI requires specialized skills and training (intelligence tradecraft). Multiple agencies contribute to this part of the mission, and their work needs to be coordinated. S&TI has traditionally been focused on assessing S&T of adversaries, using open-source and classified means. For this task, there are existent coordinators in place like the National Intelligence Manager for S&T and the National Intelligence Officer for S&T. With the rise in S&T competencies abroad and in the academic and commercial sectors, the target space for S&TI has expanded greatly.
- Creating new capabilities through the IC's own investments.

The IC's ability to adopt and leverage S&T advances and to perform S&TI are intimately linked; both call for a high priority on in-depth understanding of S&T. These two broad functions must be better integrated and coordinated across the IC. Strengthening the IC's capabilities in S&T also calls for it to take greater advantage of not only its own S&T capabilities but also those within other government agencies, across the whole of the nation, and globally. At present, the IC is not as effective as it needs to be in knowing about and agilely implementing advances that come from the global S&T ecosystem. As a result, there is a growing risk that the United States may be surprised by its strategic competitors and adversaries and may forfeit opportunities to surprise them at key times.

---

[2] Cyberspace Solarium Commission, 2020, *United States of America, Cyberspace Solarium Commission Final Report*, March, https://www.solarium.gov/report; National Security Commission on Artificial Intelligence, 2021, *Final Report: National Security Commission on Artificial Intelligence*, https://www.nscai.gov/wp-content/uploads/2021/03/Full-Report-Digital-1.pdf.

At present, the IC's connections with the broader S&T community are insufficient. One reason is security concerns that make it difficult for the IC to let potential S&T partners know about the IC's mission requirements. However, the risk of missing critical advances and being out of touch with current S&T must also be considered. It is no longer realistic to expect the United States to dominate the S&T frontier in all fields. In technical domains in which the United States is *not* the unequivocal leader, it may be preferable for the IC to engage openly with the global S&T community. Particularly in the instance of expensive, high-risk endeavors, the likelihood that IC-funded research done in isolation will be first to achieve a vital outcome is small. The greater risk is failing to be the first and then having to take significant time catching up with the rest of the world. It is the consensus view of the study committee that in some cases, the risk of not having access to the world's best researchers in a timely manner outweighs the attendant security risks of engaging with the open research community. In those cases, the security concerns can often be mitigated.

In addition, in the experience of the study committee, the IC is viewed as having in some aspects a fairly insular and risk-averse culture in regards to S&T investment and technology adoption. This often supports decisions to prioritize short-term and intermediate-term developmental activities, such as the development and implementation of S&T advances, over investing in, performing, or tracking higher-risk, often early-stage research. Early-stage research (e.g., technology readiness levels 1-3) is inherently riskier,[3] and generally takes longer to produce practical advantage. Finally, security constraints and federal acquisition practices deter many potential S&T partners (e.g., universities, multinational companies, and some non-IC federal agencies) from working with the IC. However, some changes are being made, such as the new authorities granted to CIA Labs.

These challenges were major drivers for the CSIS report mentioned above, and they apply not only to the AI-based technologies emphasized there but also to the full range of S&T. While AI, cloud computing, advanced sensors, and big data analytics (technologies emphasized in the CSIS report) will fundamentally change both the global threat landscape and the IC's tradecraft, advances from biology, chemistry, materials, quantum science, network science, social/behavioral/economic sciences, and other fields also have that potential.

In the course of conducting this study, the Committee on Leveraging the Future Research and Development Ecosystem for the Intelligence Community held two workshops and carried out a number of interviews and other exchanges with people knowledgeable about the IC's approach to S&T. (Summaries of the workshops are included in Appendixes A and B.) Those investigations, plus the depth of experience of the committee members, led to the following observations that underpin this report:

- In today's world, maintaining awareness of advances in S&T is more essential than ever, to avoid S&T surprise, to inflict surprise on adversaries, and to leverage those advances for the benefit of the nation and the IC.
- The IC, although cognizant of this need and strong in some aspects of S&T, does not give S&T the priority it merits.
- The best way to maintain awareness of S&T advances is through personal interactions between skilled IC experts and external scientists and engineers; S&T understanding is transmitted through expert networks, and not nearly as well through more passive means.
- The IC's existing efforts to track and leverage S&T need to be expanded, better coordinated, and given a higher priority.

In order to improve its capabilities for leveraging S&T, the committee concluded that the IC needs to address four major questions:

1. How can the IC agencies determine how best to spend their S&T funds, and how can those individual investments be coordinated across the IC? Better coordination—but not centralized management—would be valuable.

---

[3] Note that associated manufacturing readiness levels should be considered along with respective technology readiness levels.

2. How can the IC derive best value from—and influence as appropriate—the investments of other U.S. government agencies?
3. How can the IC gain best benefit from the efforts of U.S. industry and academia (including both in R&D and in education) and national laboratories?
4. How can the IC best interact with the global S&T enterprise? Failing to engage with the increasingly globalized S&T environment risks forgoing what may be the best available technology and raises the likelihood of technological surprise. On the other hand, there are also risks associated with adversarial access to critical S&T and on reliance on technology that adversaries originate, dominate, own, or control.

Analysis of these four broad questions is the primary focus of the report.

KEY RECOMMENDATION

The committee's analysis of the four broad questions above led to the following recommendation. This echoes the recommendation in the *Intelligence Edge* report that a chief technology officer position be established for the IC.

RECOMMENDATION 2.1: The Office of the Director of National Intelligence (ODNI) should consider elevating the priority of science and technology (S&T) by clearly designating an individual to strengthen these Intelligence Community (IC) capabilities. This individual—a chief technology and innovation officer (CTIO)—would report to the Director of National Intelligence, serve as Chief S&T Advisor to the Director, and be charged with the following responsibilities:

- Develop and maintain healthy sharing and participatory relationships across the IC and between it and many relevant domestic and global S&T entities.
- Identify S&T trends with special IC relevance and plan balanced programs of open-source and classified collection and analysis to enable their expedited development and utilization.
- Lead efforts to integrate and coordinate S&T awareness and science and technology intelligence (S&TI). Because S&TI and S&T awareness require different skill sets, and the organizational cultures endemic to each function differ considerably, the CTIO would need to be fully conscious of these differences while fostering shared capacity and understanding to benefit both enterprises.
- Convert this heightened strength in S&T to operational advantage more rapidly and agilely.
- Maintain a diverse, skilled team, selected from within the IC, to be deployed to support the above activities deemed critical to the S&TI mission.

ODNI already has a Director for S&T, but with a more limited remit and placed somewhat lower in the organization. The goal of this recommendation is to raise the visibility of S&T—assigning it to someone who reports directly to the Director of National Intelligence—and broadening the responsibilities. The committee is agnostic on whether to accomplish those goals by recasting the existing position or adding a new one. The paragraphs that follow further describe the important roles and functions of a CTIO.

This function needs to be centralized within ODNI through an office that interfaces with the global S&T enterprises, and then interfaces with the relevant IC agency S&T directorates and managers, and—separately—with those individual agency components that conduct S&TI. The purpose would be to aid those directorates and components, which would maintain primary responsibility for mission support or for S&TI. Each of these activities—using S&T for mission support, and S&TI—requires specialized skills already resident in the individual agencies, but that could be strengthened.

The CTIO should be technically astute and administratively skilled, and be given the authorities to be effective in the roles recommended here. Among those authorities, the CTIO would need to establish incentives (and reduce disincentives) for IC and non-IC entities to mutually participate in and share early results from S&T activities that may have special relevance for IC mission enhancement. The committee is not envisioning a "czar" with strong powers across the IC, but rather an individual and office specialized in managing these issues because they have become more prominent and impactful.

The exact roles, responsibilities, and authorities of the CTIO would be developed by the IC's leadership (with the approval of Congress if legislative changes are needed). The primary goal in the committee's vision is that the CTIO will need a very good understanding of what is going on across the IC: what is working well, and what is not; what shortcomings there are in the IC (or in a particular agency) regarding S&T knowledge; how external changes will affect this; and whether there are new S&T tools that could be used to improve efficiency. Based on a good understanding and process for staying current on what is happening across the IC, the CTIO would have to understand what is happening in the broad S&T community and identify areas that can help solve problems or close gaps internal to the IC. Building on this base, the office would develop and execute a process to fulfill these goals and incentivize key stakeholders to contribute to them. The committee recommends the following responsibilities:

- Outward-facing responsibilities
  — Establish and operate a coordination mechanism for S&TI carried out across the IC. This mechanism might include recruiting and assigning technically astute and operationally savvy IC S&T Liaison Officers to work closely with domestic and foreign non-IC entities, and integrate their findings and insights to maintain an awareness of the global S&T landscape.
  — Establish an S&T evaluation and assessment activity that analyzes the information gleaned from the previous item and, as needed, apprise the DNI and raise attention within the elements of the IC that are best suited to utilize the information. This activity can also feed special requests to the Liaison Officers and open-source analysts.
  — Broaden the IC's engagement with relevant FFRDCs, the Department of Defense's university affiliated research centers (UARCs), and selected academic institutions that do classified research, all of which can also provide indirect links to private industry S&T. The CTIO could facilitate partnerships and bilateral arrangements when opportunities to advance IC impact could be achieved via collaboration.
  — Expand the IC's collaborations with private corporations, non-profits, and academic institutions performing unclassified basic and applied research, including systems engineering that may have potential applications to the IC mission and pre-competitive R&D.
  — Coordinate international S&T activities across the IC and make international S&T a high priority, such as through more open engagement with allies.
- Inward-facing responsibilities
  — Serve as a high-level champion for the IC's efforts to develop or acquire and deploy transformational S&T from wherever it may arise. This could include a CTIO funding mechanism, but promising S&T capabilities relative to a particular mission area should still be managed by the agency program manager responsible for that mission and for subsequently demonstrating proof-of-concept, engineering for tailored uses, and deployment.
  — Establish and promulgate technical standards and processes to lead IC entities in transitioning S&T to reliable, high-quality operational capabilities (e.g., standards for test, evaluation, and validation including standards for independent technical review); care must be taken that this setting of standards does not become more check blocks administered through oversight.
  — Ensure that S&T best practices and S&T results (within the constraints of necessary compartmentalization) from each agency's S&T activities are shared among constituent members of the IC S&T community.

ADDITIONAL RECOMMENDATIONS

In addition to the above overarching committee recommendation, other recommendations of this report are given below, with the first digit of each one indicating the chapter in which it is found.

These recommendations are offered in a general sense ("The IC should ... "), recognizing that actual implementation would have to be assigned to specific S&T offices within ODNI and/or other IC agencies, and recognizing that, to varying degrees, some agencies are already doing some of these things. The purpose of this report is to give advice to ODNI regarding handling of S&T at its level, and not to prescribe specific actions to individual agencies. ODNI would have a special role in coordinating the implementation of these recommendations within and across the agencies, and in formulating policy regarding these recommendations.

RECOMMENDATION 2.2: The Intelligence Community should increase its ability to mine open-source science and technology (S&T) information while remaining consistent with prevailing policies and laws regarding privacy protections. It must enable integrating open-source S&T information with classified intelligence.

RECOMMENDATION 3.1: The Intelligence Community (IC) should position itself to take better advantage of opportunities afforded by interagency science and technology (S&T) committees and other contacts with non-IC agencies that have substantial S&T activities. More interagency staff exchanges would be helpful, as would more active IC participation in cross-agency research and development. Successful activities on the part of some IC agencies should be studied and mined for best practices.

RECOMMENDATION 3.2: The Intelligence Community (IC) should engage in more active partnering with Department of Energy and Department of Defense laboratories (government, federally funded research and development centers, and university affiliated research center laboratories), to take advantage of their extensive infrastructure and capabilities as well as to employ them as a vehicle for expanding the IC's access to academic and industrial research and development activities through overt relationships. The IC should increase its engagement with various laboratory program review activities.

RECOMMENDATION 4.1: The Intelligence Community (IC) should encourage its technical experts to engage more extensively on a professional level with their peers outside the relatively small IC environment. This would involve attending conferences in their respective fields of expertise, making presentations, and giving talks at other institutions, all with home agency support regarding travel, leave, and expenses. In addition, IC agency experts should be rewarded for inviting outside scientists and engineers to give talks at their home IC agencies. If the proposal to establish a chief technology and innovation officer within the Office of the Director of National Intelligence is accepted, that office would be an ideal place to encourage and oversee these practices.

RECOMMENDATION 4.2: To institutionalize increased professional interactions between Intelligence Community (IC) science and technology experts and the rest of the technical world, IC agencies should consider establishing more rotational positions for leading researchers from academia and the private sector, including start-up and venture capital communities. Because SCI level clearances often require a long approval time and impose lifelong prepublication restrictions, some of the rotational positions should be established at both the unclassified and SECRET levels. The lower security level is often subject to a quicker security clearance process than is now practical for the higher level clearances more typical of IC staff. That said, because TS/SCI clearances are usually standard for IC staff, they should be expedited when possible for rotational positions.

RECOMMENDATION 4.3: The Intelligence Community (IC) should consider emulating some of the Department of Defense's outreach efforts to scientists and engineers in research and development in order to establish trusted collaborations with academia and the private sector. Some of this is being accomplished in the IC, for example, in programs such as IC Scholarships and IC Centers for Academic Excellence. Such efforts should be expanded significantly to develop a trusted community of external researchers. This would be especially useful for engaging researchers without a long history of working with the IC.

RECOMMENDATION 4.4: The Intelligence Community (IC) should adopt more forward leaning policies for working with commercial industry to support joint IC-commercial technology development, such as data sharing, as well as acquisition approaches targeted at more effective scaling and implementation of commercial technologies, such as bridge funding. Note that data sharing should go both ways: from the industry partner to the IC agency as well. In both cases, the IC should, by working directly with industry, become part of the technology development process. Additionally, a benefit to the IC of collaboration with the private sector would be increased science and technology awareness. The IC should adopt more active policies for working with commercial industry to support joint IC-commercial technology development, such as data sharing, as well as acquisition approaches such as bridge funding targeted at more effective scaling and implementation of commercial technologies.

RECOMMENDATION 5.1: Within its mining of open-source information in general, the Intelligence Community should increase the collection of open-source information on science and technology advances and early stage companies in foreign nations. The chief technology and innovation officer could coordinate these activities and potentially assign and/or post specialists to cover key regions and countries.

RECOMMENDATION 5.2: The Intelligence Community (IC) should increase its interactions with FVEY (Five Eyes, the intelligence partnership among the United States, Canada, the United Kingdom, Australia, and New Zealand) and other allies through four steps:

1. Create a more systemic approach to cooperation, which could include having the chief technology and innovation officer develop a multi-year allied science and technology cooperation plan.
2. Set aside funding for international cooperative activities (e.g., personnel exchanges, joint research and development).
3. Support travel abroad to deepen foreign partnerships and build trusted relationships.
4. Develop common talent pools and facilitate commercial cooperation opportunities.

RECOMMENDATION 5.3: The Intelligence Community (IC) should work to establish a center (e.g., a nonprofit or at a federally funded research and development center or university affiliated research center) operated external to the IC, focused on open-source science and technology (S&T) information collection. This center should take full advantage of collection opportunities, through a presence at international symposia, where potential competitors display their state-of-the-art efforts in mission-critical areas, such as semiconductors, information technology, artificial intelligence, and machine learning, quantum computing/sensing, biotechnology, and other emergent fields of S&T.

1

Introduction

BRIEF HISTORY

In 1945, maintaining competency across science and technology (S&T) was relatively straightforward for the United States because most of the world's base for research, development, engineering, and manufacture was here. World War II had destroyed or greatly disrupted industry, education, and society in most of the rest of the industrialized world—both for victors and for vanquished. Before and during the war, many of Europe's leading scientists and engineers had sought refuge in the United States, and others would come after the cessation of hostilities, with the result that the U.S. S&T enterprise soon evolved into an attractive magnet for a disproportionate share of the world's technical talent. The U.S. wartime effort had led to the creation of an impressive array of government-funded laboratories, which were maintained, improved, and expanded in the following decades. Additionally, the creation of the national security industrial sector that operated at the technological forefront greatly enhanced our S&T dominance.

Over the decades since, the S&T landscape has become more diverse and more globally distributed. For example, the private sector in the United States now invests more in S&T than the federal government does; the commercial sector leads the defense sector in many areas; and that commercial sector is increasingly entangled with foreign partners. The rest of the world spends more on research and development (R&D) than does the United States, and the U.S. share is projected to continue declining. Although the Soviet Union is no more, Russia remains a formidable adversary in many ways, and China has emerged as our primary S&T adversary. To quote one of this report's anonymous reviewers, "China's economic growth has reached over 60 percent of U.S. Gross Domestic Product (GDP), an achievement never matched by any other country in two centuries. It is today America's primary rival in economic and scientific competition, and its ability to scale programs to competitive advantage is unparalleled. China rapidly advanced its own scientific capabilities in recent decades through focused R&D investment and by acquiring foreign technology." In addition, China is attempting to stimulate a new S&T brain drain from the United States. In the 1930s, 1940s, and beyond, scientists drifted to the United States from Europe, seeking better personal and career opportunities. In contrast, the current Chinese Sea Turtle program offers major incentives, financial and other, to U.S. scientist-citizens who were born in China, if they agree to return and help China benefit from the training and knowledge acquired in the United States.

Globalization of S&T has created opportunities for other nations as well. This democratization has been accompanied by another significant trend. Whereas the world of strategic R&D was dominated in the mid-20th century by the physical sciences and their technological applications, the past several decades have seen the emergence of information (including automation) and biological sciences[1] as major foci of research, development, and product engineering and manufacture. Each of these disciplines has its own networks and infrastructures that differ from those associated with the physical sciences. Increasingly, researchers working across technical disciplines are producing advances enabled by cross-disciplinary technological convergence.

THE CHANGING NATURE OF GLOBAL S&T

Changes in the global S&T landscape over recent decades have heightened the importance of S&T to the Intelligence Community (IC). The growth of commercial enterprises based in cutting-edge S&T—think Internet commerce, AI-based software, the bio-based economy, advances in communications, novel materials, and other developments—is evidence that technology is progressing rapidly on multiple fronts. Private-sector funding and exploitation of research have commercialized and marketed sophisticated technologies that the government has relied on for national defense. In those (many) areas in which the private sector leads the government, the government has only the same product access as adversaries and private consumers. And many of those developments are occurring not only in open-source environments such as universities and government laboratories, but also in proprietary firms.

At the same time, this increased commercialization of S&T is also becoming more international. That is especially true for fields such as life sciences R&D and computing-based technologies that have been able to take root in multiple countries. On top of these trends is China's ascension as an economic, political, and military rival, and as a leader in S&T. China's growing S&T footprint in government, military, intelligence, and commercial domains, and some positions taken by the Chinese government, combine to pose a great challenge to the United States.

The global development of increasingly sophisticated technologies creates a further challenge to the United States. The development and dissemination of such technologies—for example, cyber tools, advanced sensors, biometric devices, biomanufacturing, high-resolution imagery, enhanced technical surveillance equipment, advanced encryption, and big data analytics—means that in many cases such technologies are commercially available at a global scale. While these technological advances are enhancing our lives, they can also be innovatively used to threaten U.S. national security. The development of next-generation technologies such as the Internet of Things, 5G communications, and quantum computing will eventually present new opportunities for foreign adversaries to collect intelligence and engage in cyber operations. Technologies such as space-based collection, additive manufacturing, robotics, portable energy sources, miniaturization and nanotechnology, and advances in the life sciences, for example in synthetic biology, will also impact national security and the IC's mission.

Finally, strategic competition with China, Russia, and other countries is in many ways enabled by science and engineering. Much of the competition has so far been over technologies from AI to quantum to hypersonics. The Chinese political-economic system is well-tailored to acquiring, funding, developing, and fielding national security-relevant technologies. Indeed, China is already leading in certain technological areas that are important for national security, such as hypersonic missiles, facial recognition and space-based quantum communications. This presents both a science and technology intelligence requirement for the IC as well as a need for the IC to maintain its own competitive edge.

A component of the IC's core mission is to prevent surprise, and each of the 18 IC entities has a distinct role in that mission. An important category of surprise is S&T surprise—for example, a development in S&T by one of our adversaries that introduces a new threat to the nation's security. Thus, the IC needs to gather intelligence

---

[1] Biosciences have been at least as prominent as information sciences over the past few decades; this is especially obvious now with mRNA technology in response to the pandemic. But other areas (e.g., CRISPR, treatment of trauma, new vaccines, MRI, brain research, and so on) have arisen over the past 50 years).

INTRODUCTION

on the S&T programs of adversaries, through S&T intelligence (S&TI).[2] Beyond this, a broad awareness of S&T facilitates the IC's leveraging of the leading edge of technology for its own various missions, and to better inform U.S. decision-makers. For the IC, S&T can also serve as a tool for executing the agencies' work and as a target that adversaries might exploit to our disadvantage. A particular need is to anticipate and properly characterize potential S&T breakthroughs that could negate the effectiveness of current collection techniques and tools. This is one aspect of the "S&T as a Target" issue, and failure to correctly anticipate and prepare for any such potential changes can generate a collection gap and open the community to adversary surprise. Just a few past examples of technology shifts that presented risks to the IC include global satellite communications moving to terrestrial cable, analog moving to digital, circuit-based techniques moving to packet techniques like IP, and first-generation mobile transmissions advancing to higher capabilities. These kinds of shifts require significant preparation to deal with at scale, and thus early understanding of feasibility and trends is essential to avoid "going blind."

Successful commercial companies likewise must keep a close watch on global S&T trends, continually assess whether relative advantage/disadvantage might come from any potential S&T advances and/or competitor innovations, and adapt quickly and agilely when any such potential advantage/disadvantage becomes apparent. Failure to do so risks declining value of their products or services, global market share loss, and severe financial consequences. Similarly, the IC must have a strong understanding of how global S&T trends might impact operations. Utilization of the best available technologies to keep collection and processing, communications, and analytic processes more effective than those implemented by adversaries is clearly important.

Maintaining S&T awareness and conducting S&TI are not synonymous. The latter is focused on foreign adversaries, and it may involve results that are not accessible to the public, whereas S&T awareness is much broader, aiming to cover S&T from all sources that could be available for utilization, including mundane and open-source information, such as new patents. However, the distinction between military technology and consumer technology is blurring. Additionally, in a globalized world in which researchers routinely are educated or work across borders, in which companies are multinational, and in which research publications are instantly available around the world, the boundaries between technology created domestically, by foreign allies, or by foreign adversaries are increasingly porous. The IC must be good at both S&TI and S&T awareness, and at managing both their overlaps and necessary distinctions. This creates an additional level of complexity for the IC, which must navigate economic issues, national borders, and domestic authorities. This report of the Committee on Leveraging the Future Research and Development Ecosystem for the Intelligence Community aims to help the IC increase its capacity to navigate these converged complexities of S&TI and S&T awareness.

A recent Center for Strategic and International Studies (CSIS) report, *Maintaining the Intelligence Edge: Reimagining and Reinventing Intelligence Through Innovation*,[3] highlighted the need for the IC to relentlessly pursue competitive advantage relative to our principal adversaries, particularly by reimagining and reinventing methods to leverage technological innovation. Its focus was on the use of technological innovations that are enabled by "artificial intelligence (AI) and associated emerging technologies, including cloud computing, advanced sensors, and big data analytics." It pointed to the need for both culture change and the mastery of global S&T advances.

Building on the *Intelligence Edge* report, this report examines how the IC might better leverage the IC's internal knowledge of the full range of S&T. While the technical areas covered in the *Intelligence Edge* report will fundamentally change both the global threat landscape and the IC's tradecraft, advances from biology, chemistry, materials, quantum science, network science, social/behavioral/economic sciences, and other fields also have that potential. And in contrast to that report, which focused largely on strengthening the IC's internal capabilities with respect to AI and associated emerging technologies, this report looks across the broader government, domestic, and

---

[2] "Scientific and Technical Intelligence (also Science and Technology Intelligence or Scientific and Technological Intelligence; abbreviated S&TI) is the systematic study and analysis of foreign capabilities in basic and applied research and applied engineering. S&TI products are used to warn of foreign technical developments and capabilities and to guide the development of future capabilities, which are often provided through R&D." See United States Intelligence Community, 2013, *Report of the National Commission for the Review of the Research and Development Programs of the United States Intelligence Community*, Unclassified Version, https://www.intelligence.senate.gov/sites/default/files/commission_report.pdf.

[3] Center for Strategic and International Studies (CSIS), 2021, *Maintaining the Intelligence Edge: Reimagining and Reinventing Intelligence Through Innovation*, a report of the CSIS Technology and Intelligence Task Force, Washington, DC, January.

global S&T environments, taking into account how those environments are evolving.[4] The Chapter 4 section titled "Dynamic Nature of Today's S&T Ecosystem" provides one example that extends the lessons from the CSIS report.

The committee generally concurs with the major findings of the CSIS report, and its recommendations are consistent with them. In particular, both reports recommend elevating the focus on S&T by creating a "C-suite" position, either chief technology officer or chief technology and innovation officer. Going one step further than the CSIS report, this report argues that S&TI and S&T awareness are necessary core components of intelligence in the great power competition as a means to avert strategic S&T surprise and to provide the IC with an S&T edge against competitors;[5] and can also provide insight into reducing the loss of U.S. intellectual property to others. This report makes recommendations regarding how the IC can (1) innovate and leverage advances from other fields of S&T and (2) be better positioned to identify, track, and employ developments across all of S&T in the service of the core intelligence missions of collection, analysis, and distribution. Because the needs for S&T awareness vary greatly across the IC mission areas and agencies, how best to strengthen that capability is not amenable to a one-size-fits-all approach. For that reason, this report leaves the details to the IC, to be worked out in classified deliberations. What is clear, however, is that any proposed approach for strengthening S&T awareness must be cognizant of current and future gaps and be tailored to each of the cultures and sub-cultures across the IC. This could not be addressed in this unclassified study.

WHAT DOES IT MEAN FOR THE IC TO "LEVERAGE" THE S&T LANDSCAPE?

Although the R&D landscape is evolving, the IC's core mission remains the same—to understand and anticipate global trends and to deliver timely, accurate, and relevant analysis of those trends to U.S. policy- and decision-makers so as to avert strategic surprise and provide those leaders with decision-advantage over competitors. Advances in S&T improve our options for intelligence collection, as well as for the IC's own internal operations. And the impacts of nontraditional threat areas such as climate change (with geopolitical impacts that can affect both diplomatic and military planning) continue to expand the range of S&T that the IC must monitor and evaluate.

Currently, the IC is not well positioned to leverage the rapid developments in S&T of recent years and anticipated in the future. When a potentially game-changing insight originates somewhere in the S&T ecosystem, someone in the IC not only has to recognize the potential for where it might be advantageously applied, but also must have the bureaucratic skill to assemble adequate funding and staff for rapid implementation. At present, the IC's capabilities for recognizing game-changing S&T developments and their implications for intelligence operations, and rapidly converting them to operational advantage, are not as efficient or effective as they need to be.

The IC must therefore have a comprehensive, global awareness of, and access to, a broad range of S&T, including S&T that is still under development. A foundational strategy exists to support this. For example, the National Security Strategy (NSS) of the United States of America (December 2017) directs the IC to master strategic intelligence issues through research, knowledge development, collaboration with experts within the IC, and outreach to experts in academia and industry, as well as the use of advanced analytics and tradecraft, to provide assessments on the strategic context for the policy- and decision-makers.[6] The IC's overall S&T strategy is articulated in the fiscal year (FY) 2016-2020 IC S&T Strategic Plan, which has the objective of ensuring future

---

[4] In fact, the *Intelligence Edge* report (footnote 3) recommended that ODNI, "in partnership with the National Academy of Sciences, should sponsor a study on the potential intelligence collection applications and implications of synthetic biology and associated technologies." The CSIS report mentioned as just one example that "synthetic biology, its convergence with AI and computational power, and possible intelligence mission applications such as biosurveillance applications and remote sensing with biological systems [stand] out as having a potentially transformational and generational impact for the collection of intelligence."

[5] See G.C. Allen, 2019, *Understanding China's AI Strategy*, Washington, DC: Center for a New American Security, posted February 6, http://www.cnas.org/publications/reports/understanding-chinas-ai-strategy. Strengthening the discussion in footnote 3, above, again from the *Intelligence Edge* report, there is a particular challenge from China, arguably the strongest U.S. competitor in developing key technologies. Chinese leadership is reportedly convinced that AI technology is critical to global power competition, both militarily and economically.

[6] White House, 2017, *National Security Strategy of the United States of America*, December, Washington, DC, p. 8, https://trumpwhitehouse.archives.gov/wp-content/uploads/2017/12/NSS-Final-12-18-2017-0905.pdf.

intelligence advantage and guiding the application of S&T to solve intelligence problems.[7] Also of relevance, the Office of the Director of National Intelligence's (ODNI's) Director of S&T developed the Intelligence Science and Technology Partnership (In-STeP) as the foundational mechanism and process structure to assist the IC's S&T investment decisions.[8] In-STeP facilitates this by providing information to support greater synergy in the intelligence-related research efforts of the S&T enterprise, the broader U.S. government, foreign allies, partners in industry, FFRDCs, and academia.[9]

Coordination of the IC's S&T efforts is managed on an ongoing basis through (1) ODNI's Director for S&T (D/S&T), and (2) the National Intelligence S&T Committee (NISTC), which is chaired by the D/S&T. The NISTC is composed of the principal science officers of the National Intelligence Program, and one of its mandates is to coordinate advances in research and development related to intelligence. This report focuses on how the IC can enhance its access to the broad S&T landscape. It considers that landscape as consisting of four general domains: S&T expertise and activity within the IC itself; expertise and activity within the government more generally; the U.S. base of S&T activities; and the global enterprise.

The IC currently faces several factors that impede its ability to participate in such a global, collaborative ecosystem. First, IC security constraints prevent the IC from sharing many of its authentic concerns and insights with potential S&T partners. In addition, based on its extensive interviews and its members' knowledge, the committee concludes that the IC is widely viewed as having a fairly insular and risk-averse culture, at least with respect to S&T investment and technology adoption, which inhibits the appetite to collaborate with outsiders. Finally, the combination of security constraints and government acquisition practices act to discourage potential S&T partners (e.g., universities, multinational companies, non-governmental organizations, and some federal agencies from outside the IC) from working with the IC. These factors are well recognized. For example, the IC's FY 2016-2020 S&T Strategic Plan observes that it can be challenging for the IC to work with the basic research community because access to classified materials, security clearances, reporting requirements, and travel restrictions imposed on researchers may be perceived as inconsistent with the open, international nature of basic research.[10] It identified the need for the IC S&T enterprise to be a "selectively permeable information conduit" between the open and classified worlds while staying ahead of security challenges and foreign counterintelligence activities.[11]

In public discussions, several senior officials have made reference to these problems. For example, during an October 2021 announcement of Central Intelligence Agency (CIA) organizational changes, Director William J. Burns noted that CIA has long talked about the need to increase its talent pool. However, the long process to hire a candidate is often a deterrent. In order to attract and hire talent, Burns announced changes to processes to significantly reduce the time it takes for applicants to join CIA. He also announced the launch of the CIA Technology Fellows program to bring promising experts to CIA for 1 to 2 years of public service.[12]

Some of the individual IC agencies—particularly the "big five"[13]—have well-funded, well-managed S&T programs to support their individual missions. However, those programs quite naturally tend to prioritize short- and intermediate-term developmental activities over a broader, or longer-term, S&T awareness. The IC as a whole could benefit from adding complementary awareness of S&T that might not link to near-term missions. This prioritizing of near-term missions, when funding S&T projects, is especially problematic if needed R&D is to be performed in an academic setting (see Chapter 4). Without such an overview capability, individual agencies could be limited in their abilities to identify threats and opportunities outside their established scope. Efforts to coordinate and to provide a broad view exist within ODNI, but these have been limited in scope.

---

[7] Office of the Director of National Intelligence (ODNI), 2016, *The FY 2016-2020 IC S&T Strategic Plan: Managing Risk to Ensure Intelligence Advantage*, Washington, DC: ODNI S&T, https://www.dni.gov/files/documents/atf/In-STeP%20-%20Strategic%20Plan.pdf. ODNI is currently developing a new Strategic Plan for FY 2021-2025.

[8] ODNI, 2016, *The FY 2016-2020 IC S&T Strategic Plan*, p. 7.

[9] ODNI, 2016, *The FY 2016-2020 IC S&T Strategic Plan*, pp. 7 and 27.

[10] ODNI, 2016, *The FY 2016-2020 IC S&T Strategic Plan*, p. 3.

[11] ODNI, 2016, *The FY 2016-2020 IC S&T Strategic Plan*, p. 3.

[12] J. Hamilton, 2021, "CIA Wants to Speed Up Hiring Process and Leidos Wins $300M NSA SIGINT Award," October 8, https://news.clearancejobs.com/2021/10/08/cia-wants-to-speed-up-hiring-process-and-leidos-wins-300m-nsa-sigint-award.

[13] The Central Intelligence Agency (CIA), the Defense Intelligence Agency, the National Security Agency (NSA), the National Geospatial-Intelligence Agency (NGA), and the National Reconnaissance Office (NRO) are often referred to as the "Big Five," within the IC.

CHARGE FOR THIS STUDY

The study that led to this report was commissioned by ODNI, which defined the following charge:

> The National Academies of Sciences, Engineering, and Medicine will appoint an ad hoc committee to plan, organize, and write a consensus report that will be based largely on two unclassified workshops. The first workshop will review projections of the evolution of the R&D ecosystem and implications for the Intelligence Community (IC). The second workshop will focus on how elements of the IC might strengthen and enhance their ability to contribute to the maintenance of the R&D ecosystem to help ensure that the R&D needs of the IC are included in the R&D agenda, leverage that system, and benefit from the investments of other government agencies and the private sector. Following each of the two workshops, an NRC-appointed rapporteur will produce a workshop proceedings-in-brief that will be provided to the sponsor.
>
> After reviewing the workshop results and consulting with the sponsor, the study committee will hold meetings to gather further information and perspectives, and to hold interactive substantive discussions. The study committee will produce a peer-reviewed consensus report, including findings and recommendations. In deliberating and writing its report, the committee may also address topics such as:
>
> - What are basic perceptions of the goals of the nation's R&D ecosystem?
> - What are current approaches for ensuring a healthy R&D ecosystem?
> - Given twenty-first century realities, what are realistic goals and expectations?
> - How might the US IC work to facilitate a whole-of-government approach for a robust innovation ecosystem?
>
> The project will produce (1) two unclassified, workshop proceedings-in-brief prepared by a designated rapporteur; and (2) an unclassified, peer-reviewed consensus report that draws from the information gathered during the workshops and other information sources, as needed, and will provide findings and recommendations. If cleared members of the committee believe that the work warrants a classified annex to the unclassified report, the sponsor will accept that as part of this task.

Because the four bulleted questions in this charge are optional ("the committee may also address"), the committee did not attempt an in-depth answer to the first three of them. It focused its attention on the final bullet, examining it through four related questions, with the concurrence of the sponsor:

1. How can the IC agencies determine how best to spend their S&T funds, and how can those individual investments be coordinated across the IC? Better coordination—but not centralized management—would be valuable.
2. How can the IC derive best value from—and influence as appropriate—the investments of other U.S. government agencies?
3. How can the IC gain best benefit from the efforts of U.S. industry, academia (both in R&D and education), and national laboratories?
4. How can the IC best interact with the global S&T enterprise? Failing to engage with the increasingly globalized S&T environment risks forgoing what may be the best available technology and raises the likelihood of technological surprise. On the other hand, there are also risks associated with adversarial access to critical S&T and on reliance on technology that adversaries originate, dominate, own, or control.

In meeting this charge, the committee considered direct investments, partnerships, venues for cooperation and coordination, applicable laws and regulations (both domestic and international), and mechanisms for setting standards. This is, in large part, a study of how the IC can best employ its assets to appropriately leverage the broad S&T enterprise, including through S&T funding, through participation in interagency forums, and by strengthening relationships with universities, corporations, and related organizations. The charge suggests that the IC already recognizes a need to strengthen its leveraging of S&T, and is seeking advice on options for doing so. Therefore, the committee did not attempt to make that case, although it did draw on its own experience and information gathering to prioritize its suggestions.

HOW THE STUDY WAS CONDUCTED

The study charge is very broad, but it divides into four general questions related to domains in which IC-relevant S&T takes place: within the IC, elsewhere in the federal government, within domestic academic and commercial entities, and internationally. The committee notes that, while those charge questions may be clearly stated, they are not cleanly separated, and significant overlap was to be expected. The committee also agreed that the intent of the study (and report) would be to identify major issues, and present associated findings and actionable recommendations to support the overall goal of getting best access to the S&T landscape and to use it to full advantage. Most of the report's recommendations are offered in a general sense ("The IC should ... "), recognizing that actual implementation would have to be assigned to specific S&T offices within ODNI and/or other IC agencies, and recognizing that, to varying degrees, some agencies are already doing some of these things. The purpose of this report is to give advice to ODNI regarding handling of S&T at its coordinating level, and not to prescribe specific actions to individual agencies. ODNI has a special role in coordinating the implementation of these recommendations within and across the agencies, and in formulating policy regarding these recommendations.

To address this charge, the committee focused on the four domains of S&T mentioned above. Chapters 2 through 5 address these questions in series, gradually opening the aperture to consider broader swaths of the overall S&T enterprise. To inform its deliberations, the committee held discussions with over 20 subject-matter experts during its meetings, and organized two information-gathering workshops. Summaries of those valuable workshops are included in Appendixes A and B. In order to gain an understanding of how R&D is managed within IC agencies, the committee's staff conducted interviews with several S&T managers. The results of those interviews were determined to be controlled unclassified information (CUI) and were made available to committee members through appropriate channels. These interviews provided insights into how the S&T managers manage their programs, cooperate and partner outside their agencies, and coordinate across the IC, and handle some personnel matters. In addition, this report draws on the extensive experience of the members of the study committee (whose biographical information is provided in Appendix D).

Each chapter is generally organized around the following topics: (1) What does the IC need to be doing? (2) What are the current impediments to accomplishing those tasks? (3) What does the committee recommend to overcome these impediments? Since the scope of the problem varies across the four domains, the chapters are organized similarly, but not identically.

This study is unclassified, so the committee did not have access to information regarding specific IC S&T projects and budgets or regarding near-term or future gaps and challenges.

2

A Vision for Strengthening the IC's Ability to Leverage S&T

Based on the personal experience of the members of the study committee, interviews, discussions at the two workshops held to inform the committee (see Appendixes A and B), and other information gathering sessions and literature review, the committee finds that elements of the current Intelligence Community (IC) science and technology (S&T) enterprise are not optimized to leverage the evolving S&T landscape within and beyond the IC. This finding echoes the 2021 Center for Strategic and International Studies (CSIS) *Intelligence Edge* report cited in Chapter 1, which offered a large number of detailed recommendations to help the IC strengthen its capabilities in S&T.[1] The committee agrees with many of those recommendations and has attempted to cover complementary ground. This chapter begins with a brief discussion of how the IC currently works to stay abreast of the broad S&T enterprise, without repeating the content of that CSIS report. It then relates some particular challenges that limit those efforts, and then offers suggestions for strengthening the IC's capabilities for monitoring and leveraging S&T advances.

EXISTING IC LEVERAGING OF S&T

Each IC agency's research and development (R&D) investments and associated S&T activities are focused primarily on its own mission space. This includes supporting the discovery and development of S&T that it can use, maintaining sufficient S&T awareness, and contributing to the collection of S&T intelligence (S&TI). Agencies can, and do, coordinate and collaborate on S&T areas of mutual interest.

The Office of the Director of National Intelligence (ODNI), particularly the ODNI Director for S&T (D/S&T), has a mission to facilitate this coordination and develop IC-wide S&T strategy. D/S&T is focused on the "blue" side—that is, enhancing U.S. capabilities to take advantage of S&T advances. Also within ODNI are the National Intelligence Officer (NIO) for S&T and the National Intelligence Manager (NIM) for S&T. The offices for S&T within NIO and NIM are smaller than that of D/S&T, and they report to a different Deputy DNI (the DDNI for Mission Integration) than does D/S&T (which reports to the DDNI for Policy and Capabilities). NIO and NIM have broad interactions across the IC. Their "red" side insight is important for informing and shaping the U.S. S&T program. It is also extremely useful in keeping both our collection and covert action capabilities effective. The split responsibilities of D/S&T, NIO, and NIM create some unhelpful compartmentalization that limits the

---

[1] Center for Strategic and International Studies (CSIS), 2021, *Maintaining the Intelligence Edge: Reimagining and Reinventing Intelligence Through Innovation*, a report of the CSIS Technology and Intelligence Task Force, Washington, DC, January.

IC's ability to build and maintain a coherent, integrated view of the overall S&T enterprise. In addition, both red and blue perspectives have a shared need to understand global S&T activities.

Each agency has an S&T organization that has evolved according to the needs identified by that agency's director and senior management. This is a consequence of each agency having its own history and its own R&D culture, which predates the establishment of ODNI. The agencies differ in matters discussed in this report, such as (1) balancing short- and long-term research; (2) collaboration with other agencies; (3) maintenance of in-house R&D capabilities; (4) assimilating relevant S&T advances; (5) making use of commercial capabilities and sources; (6) leveraging the technologies of major defense and aerospace systems integrators; (7) interacting with small innovative companies and other such innovators; and (8) interacting with the academic community. These differences affect how the recommendations of this report should apply to each individual IC agency.

Each of the "big five" has an S&T director charged with managing the agency's R&D investments and related activities. In addition to their differences noted in the previous paragraph, the agencies' S&T, their S&T structures have been evolving, and are likely to continue to do so. For example, in 2021, the CIA announced several major initiatives, including the establishment of a chief technology officer (CTO), a Transnational and Technology Mission Center, a Technology Fellows Program, and the CIA Labs organization.[2]

The S&T managers and directors across the IC agencies are charged with supporting their agency's mission programs and they generally have wide latitude in deciding how to allocate funding and personnel to meet their missions' current and future needs. Specifically, S&T managers in at least the major agencies view mission support as calling for a mixture of funding R&D to build near-term capabilities, future-oriented research in anticipation of the next-generations of current mission systems, and more speculative, higher risk, investments aimed at finding new ideas to support future missions. Investments can be in-house, at or with other agencies, or contracted out to academia, industry, and other performers. The amount of in-house R&D capabilities varies widely across agencies. One agency S&T manager characterized that agency's in-house effort as being like an industrial research laboratory in size and scope, while others commented on having very little in-house R&D performance capacity.

While the mix of investments across these categories is under the control of the S&T managers, their options are subject to basic constraints, some of which can be substantial and place significant limitations on the flexibility of allocations within some spending categories. The S&T managers need to keep the agency mission managers—and ultimately the agency directors—satisfied that they are being adequately supported. The way money is appropriated limits discretion regarding allocation across R&D categories (e.g., basic research, applied research, advanced development). The funding is generally weighted toward higher technology readiness levels (TRLs)—that is, toward technologies that are moving close to implementation.

Investments are not made in isolation, and often they are coordinated with partners with similar interests at other IC agencies, non-IC government agencies, and other partners (including foreign partners). There is increasing interest in establishing research efforts that are more open to, and in contact with, the external S&T community, and the study committee believes this is both healthy and necessary. The IC's ability to attract S&T talent benefits from increased interactions with the R&D sectors in both academia and industry, at both the student and mid-career professional levels. Such interactions could include simple collaborations as well as personnel exchanges. Such rotational assignments are proven methods for enhancing technical knowledge and skills.

NSA provides an excellent example of utilizing a diverse set of partners to meet S&T needs, including in-house research, federally funded research and development centers (FFRDCs) and (the Department of Defense's [DoD's]) university affiliated research centers (UARCs). One of these is the University of Maryland's Applied Research Laboratory for Intelligence and Security, the only UARC focused on IC research. NSA also funds the Laboratory for Analytical Sciences at North Carolina State University.

The CIA is in the second year of standing up its new federal laboratory, an outward-looking organization that supports unclassified research and open interactions with researchers in academia and industry. "CIA Labs conducts multidisciplinary research, development, testing, and engineering to address new challenges; adapt, improve, or accelerate the production of existing solutions; and solve persistent scientific and technological problems in new

---

[2] O. Gazis, 2021, "CIA Creates New Mission Centers Focused on China and Technology," *CBS News*, October 7, https://www.cbsnews.com/news/cia-creates-new-mission-centers-china-and-technology.

ways."[3] CIA Labs also intends to bring intelligence officers, academia, and private-sector companies together to develop technologies useful to the mission of the agency. CIA Labs will likely put the CIA and the rest of the IC in a stronger position to utilize the skills and creativity of academia and the private sector in accomplishing their missions.

Other IC agencies are considering establishing open laboratories. One objective of open laboratories is to provide a clean separation between unclassified research and classified work. Depending on the results, some information developed through unclassified work—which can draw on the world's best scientists and engineers, and with the free flow of information allowing much more rapid advances than in classified research—can be brought "in-house" for further development as a classified project. Those working on the unclassified work are not subject to the restrictions associated with classified work, and are free to continue other open research. While this has its attractions, some of the IC managers with which the committee interacted cited security concerns with this approach, due to the proximity of cleared and uncleared individuals working closely on the same or related projects.

S&T managers within the IC also have several options for expanding the career experience of their S&T staffs. These options include rotating personnel among in-house positions; bringing in outside R&D talent on temporary assignment through the Intergovernmental Personnel Act Mobility Program (commonly known as "IPAs"); sending agency personnel out on temporary assignments; and sharing staff through joint duty assignments (JDAs). Such arrangements build relationships across agencies. One S&T manager reported that typically 20 percent of staff are on JDA at any given time. In fact, some agencies require JDAs or other rotations outside of the home agency when considering career advancements. In addition, S&T personnel are often encouraged to take assignments within their own agencies in units outside of S&T.

Some of the managers interviewed send personnel on temporary assignments outside the government, such as to academia or industry, and encourage senior academic researchers to take sabbaticals at the agency or otherwise become agency subject-matter experts. However, career rotations are not necessarily easy for the employee. Scientists and engineers from outside the IC do not necessarily have sufficient understanding of the intelligence profession and culture; and intelligence officers do not necessarily understand the world of science.

In addition to individual broadening assignments, the S&T managers interviewed by the committee staff noted important benefits from establishing R&D partnerships with entities outside their agency. Partnering must, of course, be done consistent with necessary security, and satisfy other such constraints. In the view of these IC S&T managers, partnering expands opportunities to meet mission needs, and it also helps the agency look beyond current mission needs to anticipate future mission needs, thus broadening the agency's relevant situational awareness and enhance the country's ability to leverage these non-IC S&T investments. Moreover, partnering arrangements provide a vehicle for personnel exchanges, rotations, and other forms of temporary assignments.

IC agencies partner with other IC agencies, with selected other U.S. government agencies, with the private sector, and with foreign partners (mostly Five Eyes[4] partners, NATO, and other allied governments). Some of these partnering arrangements are formalized through activities such as regularly scheduled meetings; others are on a more ad hoc basis. One manager noted that the United States has become skilled at the mechanics of collaboration, but is not so good as a member of a collaborative team. Another suggested that it would be worthwhile to study how other entities (e.g., large corporations) partner in S&T.[5]

Based on the committee's interviews, many S&T managers in the IC appear to agree that the National Intelligence Science and Technology Committee (NISTC) has not been as useful as it could be. While NISTC meetings provide somewhat of a venue for portfolio reviews, they do not offer sufficient time to do this in depth. Interviewees used terms like "general awareness" and "skim the surface" to describe the level of depth. Also, because NISTC connects only the top-level S&T managers, it does not promote working relationships among S&T experts from different agencies. One manager suggested that NISTC could be more useful if it were to provide a "matchmaking" function to assist agencies in matching needs, assets, and solutions. NISTC is now being revived,

---

[3] Central Intelligence Agency, "CIA Labs," https://www.cia.gov/cia-labs, accessed August 7, 2021.

[4] Office of the Director of National Intelligence, "Five Eyes Intelligence Oversight and Review Council (FIORC)," https://www.dni.gov/index.php/ncsc-how-we-work/217-about/organization/icig-pages/2660-icig-fiorc, accessed January 26, 2022.

[5] Benchmarking best practices might be analyzed in some detail, but this would imply a more detailed study.

with adjustments such as data calls and associated databases, and some interviewees feel it is becoming much more useful than it had been.

ASPECTS THAT LIMIT IC CAPABILITIES FOR LEVERAGING S&T

As noted in Chapter 1, the IC faces several major challenges that impede its efforts to maximize its ability to assess and access the evolving S&T landscape—including culture, security needs, and government practices. The committee examined these impediments, and found them to be credible. In particular:

- While most R&D in other contexts is conducted in an environment of open collaboration, the IC culture is generally closed, primarily due to security considerations.[6] Outside the IC, S&T is a global enterprise, and academic and industry teams based in the United States typically have members who are not U.S. citizens. The IC does not have the same freedom to partner with its rivals that some commercial entities do, and it must generally be cautious with interactions that might involve non-citizens and/or be open to all in the S&T enterprise. Thus, the IC's access to the global S&T landscape is necessarily constrained.
- The research and technology environment routinely operates within a digital architecture with increasing global access to data, information, technologies, and enhanced computing capabilities, and as a result, progress often advances very rapidly. Government rules, regulations, and procedures tend to restrict the agility of government agencies to follow the advances, and IC agencies typically are even more encumbered. Procurement and contracting requirements can be very time-consuming and too restrictive for the rapid pace of discovery and technology. For example, the pace of refresh of office information equipment within the IC is often much slower than it is in the commercial world, and connections to open networks are often difficult or impossible. One of the issues that is increasingly slowing down the acquisition of new office technology is the Internet of Things (IoT) challenge. A lot of new equipment comes with integrated Bluetooth and WiFi. This is a huge security issue for secure facilities. The deeper these radio frequency (RF) components are buried, the harder it is to remove them and/or to ensure that they cannot be used to compromise the secure facility.
- Committee interviews with IC S&T managers revealed that, while some innovative approaches are available, contracting rules and procedures are still viewed as insufficiently flexible to meet the needs. The process of contracting for R&D is often perceived as being too long to meet needs for timely interaction, too complicated for those who do not already have a history of working with the government, and too restrictive.
- The government's restrictive personnel practices impede the IC's abilities to employ experts in rapidly emerging technical disciplines. Salaries are often significantly below commercial or academic competition, hiring decisions move slowly, and long delays for matters such as security investigations can stretch the "on-boarding" process out for many months or even years. In addition, recruiting fresh talent can be hindered by necessary restrictions on foreign travel and foreign contacts, limitations on digital social activities, and by lifetime prepublication review requirements for those who are granted certain clearances. The latter is especially problematic for academic researchers who might otherwise be recruited for a temporary position at an IC agency.
- Achieving striking advances in research often requires the acceptance of risk and failure, and the IC tends to be risk-averse. In order to work at the leading edge of research, the IC needs to have greater tolerance for failure in the early stages of high-risk, high-reward research projects. To their credit, IC agencies do fund some cutting-edge work. Being involved in cutting-edge R&D is a time-tested way for an enterprise to maintain a high level of awareness and insight. However, it is not possible for the IC to be directly engaged in all relevant areas of S&T, so supplementary means of gaining awareness and insight are also needed.

---

[6] Including controlling important information in compartments with very limited access.

- As IC responsibilities are expanded into non-traditional areas such as climate change, global health threats, and disinformation campaigns, IC S&T is not adequately resourced and prepared for the additional demands of these new tasks. For reasons such as those discussed above, it is not feasible for the IC to have in-house capabilities in all of these relevant disciplines, and token involvement is not generally adequate to provide awareness and insight.
- IC agency S&T programs tend to be tightly "stovepiped" to meet agency mission needs, and to protect compartmentalization of highly classified information. This narrow focus conflicts with the broad perspective necessary to discern and follow advances in a timely way, so proactive steps to foster sharing and coordination are necessary.
- Finally, IC S&T is not configured to take a broad, integrated approach.[7] While ODNI does have a Director of Science and Technology who serves as a focus for S&T monitoring and assessments for that office, the position is not configured or resourced to be able to fully understand the needs of the IC S&T community, coordinate S&T activities, and maintain broad awareness of the national and internal S&T landscape. Under existing authority, the role of D/S&T is more coordinator than leader among IC agencies.

FOUNDATIONAL STEPS FOR STRENGTHENING IC S&T CAPABILITIES

Improve Coordination and Collaboration Among the IC Agencies

The ability of individual agencies to share knowledge regarding S&T development work, their mission needs, and technical solutions is currently limited by a combination of security constraints (plus an internal culture that is cautious about sharing) and the absence of incentives for more collaboration. Interviews with S&T managers produced observations that more encouragement and rewards for communications among agencies are required; lack of interagency communication (on S&T) is a significant problem; and that discussions on S&T matters at NISTC meetings need to be more comprehensive. Some interviewees noted that D/S&T has made important progress in improving communication. A truly coordinated S&T system would understand and evaluate what S&T tools are needed by the various IC agencies to perform their missions; assess the availability of such tools; promote the sharing of S&T information, expertise, and technologies among the IC agencies; strengthen networks among the IC S&T managers to support their missions by assuring systems development of needed capabilities through operation availability; and strengthen the collective capabilities of S&T intelligence.[8] Interviews with IC S&T managers and the relevant experience of committee members has persuaded the committee that these objectives are not being attained as effectively as they could be.

Strengthen the IC's Capabilities for Monitoring All of the Relevant S&T Landscape

Because advances in S&T increasingly offer opportunities for the IC to increase mission effectiveness and impact—and also recognize risks of surprise—it is urgent that the IC develop a stronger situational awareness of S&T across the globe. It must create or expand its abilities to liaise with non-IC scientific and technical experts who can provide the IC with open-source situational awareness of the increasingly rapid S&T developments in areas of interest. These networks of contacts will link the IC to experts from U.S. research institutions and corporations, from within and outside the government, as well as from international components of the S&T enterprise. The purpose of these contacts would be on collecting, mining, and assessing open-source S&T information,[9] and commensurate capabilities must be established in order for this enhanced understanding to be shared across the

---

[7] For example, it may be impossible for breakthroughs or special insights known to one compartment to be shared with other compartments.

[8] As noted in Chapter 1, science and technology intelligence (S&TI) is the systematic study and analysis of foreign capabilities in basic and applied research and applied engineering. The emphasis within S&TI work is on capabilities and advances that are not documented in the open literature, especially those arising in other nations, whereas most mentions of "S&T" in this report refer to a broader range of science and technology, including open-source information. Regular meetings among appropriately cleared individuals from the various agencies would be highly beneficial.

[9] Where appropriate, cleared individuals within these networks could advise the IC on classified S&T progress around the world.

IC in a way that facilitates its ultimate use. This topic was addressed extensively at the workshops. For example, one speaker championed drawing not only on the competitive intelligence of academia but also on the competitive intelligence of the commercial sector. The challenge in both cases is that it can be difficult for the IC to reach out directly to those universities and companies without a contract in place (i.e., multiple layers of approval are needed). Discussions of how IC outreach to academia and the private sector might be accomplished are noted in corresponding sections in Chapter 4.

This speaker also suggested creating a clearinghouse that is separate from but accessible to the federal government to facilitate interactions with these academic and commercial communities, which could lead to more timely, cost-effective, and accurate awareness of global S&T. The Director of the Office of Naval Research (ONR) Global described how it works internationally, establishing trusted partnerships with both other U.S. agencies and allied intelligence agencies to understand state of S&T in a broad spectrum of other countries. Similarly, it was pointed out that "In December 2020, *Department of Defense International Science and Technology Engagement Strategy: A Unified Approach to Strengthen Alliances and Attract New Partners* (ISTES) was released, providing a rationale for engaging with other countries. The mission of ISTES is to leverage foreign defense S&T capabilities, develop relationships with other countries to access these capabilities, maximize coalition interoperability, and achieve U.S. national security objectives."

Increased networking would augment the IC's current ability to maintain awareness of S&T developments within and outside the United States, and foster partnerships with non-IC entities. This could be accomplished through greater engagement in S&T conferences around the globe as noted above or, for example, funding emergent start-ups, both domestically and internationally. The effort should supplement, rather than supplant, the networking activities conducted already by IC S&T offices. Such an outward-looking effort could be staffed by scientists or engineers from the IC or on assignment from other federal agencies, federal laboratories, academia, or industry, in a manner that learns from the experience of ONR's Global program, through which scientists and engineers with experience conducting R&D (rather than intelligence officers) handle many of the day-to-day interfaces with the non-IC community. That model has enabled ONR to build trusted S&T relationships with researchers around the globe.

Improve Methods for Collecting and Using Open-Source Information, with Emphasis on S&T

To enable this S&T situational awareness, the IC needs the technological tools to collect, manage and exploit enormous quantities of S&T data, both from open and classified sources. In today's environment, such data can provide deep insights into other nations' S&T activities and capabilities. Open-source intelligence is particularly germane to S&T awareness because so many of the advances in S&T take place outside of governmental efforts. Moreover, unlike much—and probably most—of the information available from open sources, S&T information is vetted for accuracy and reliability through an open process of peer review and competitive research. (It is important to note, though, that an increasing amount of information is being communicated in advance of formal publication in peer-reviewed journals, via preprints and other informal mechanisms.) Information collected from sources such as scientific publications, professional society conferences, and patent applications is inherently more credible than information from press reports, social media, and commercial data merchants.

From a more general intelligence perspective (i.e., beyond gathering information on S&T activities), the potential value of open-source intelligence (OSINT) has been recognized for decades, and efforts have been mounted to facilitate such work. For example, in the past, the CIA operated the Open Source Center, which was supplanted by the Open Source Enterprise. However, it has been challenging for the IC to mine this enormous quantity of data, identify and select the data of most significance to the IC, verify the reliability of the open-source information, and integrate it with classified sources to produce analytical products. There has been continuing progress in addressing this integration, applying advanced analytics to the combined data, while also respecting privacy issues. It would be beneficial to advance these efforts to satisfy the need to combine classified and open-source data in order to derive reliable information from both types of sources.

In addition to continuing to conduct its critical mission of collecting and analyzing information from clandestine sources, the IC also needs to more fully recognize the value of open-source S&T information, particularly to support S&TI and S&T awareness, and further its ability to collect, analyze, and integrate S&T information from open sources.

Overcome Bureaucratic Obstacles That Impede IC Efforts to Fulfill Its S&T Missions

The seven items discussed in the section above on "Aspects That Limit IC Capabilities for Leveraging S&T" suggest that IC agencies face some common bureaucratic challenges with respect to fully using the IC's S&T capabilities. A shared approach to addressing them would be more efficient and effective than separate efforts. An example of such a shared challenge is the systemic barriers in hiring and retaining S&T professionals. These barriers include requirements for security clearances and the lengthy clearance process, government salary scales, impediments to career path opportunities, and restrictions on publication and on travel.

IC access to, and understanding of, the broad national and global dimensions of the S&T landscape requires IC personnel who interact with professionals across that landscape. Timely awareness of and access to new developments requires the IC to recruit and retain scientists and engineers who are current in their fields and have active working contacts with those conducting leading edge R&D. Thus, the IC needs to develop personnel policies and practices that attract qualified people to the IC in a timely manner. In the experience of the study committee, reinforced by comments received during its interviews and two workshops, the IC could do more to improve its hiring and retention of S&T professionals.

Another shared bureaucratic barrier that limits the IC's capabilities with S&T arises from contracting processes that hamper access to rapidly developing technology. Innovation is often stifled by inappropriately detailed contracting requirements. Lengthy contracting processes are a broad problem within the IC, particularly for rapidly evolving technologies such as computing capabilities. In addition, there are apparent cultural preferences toward risk avoidance in developing innovative contracting procedures and away from use of less-common contracting approaches. Congress has authorized DoD to run a pilot program to explore novel contracting for software-related initiatives, and some agencies have sought to exploit other transaction authorities to allow them to be more innovative with their contracting. Shared attention to these options, and to minimizing cultural limitations, could facilitate more nimble contracting.

A third shared bureaucratic barrier arises from requirements related to International Traffic in Arms Regulations (ITAR), which were designed to control dissemination of information related to weapons systems and military technologies to all nations. At present, these regulations can impede timely S&T collaboration with NATO, the Five Eyes intelligence alliance, and other allies and thus unduly constrain the IC's work. The intent of ITAR is to safeguard sensitive U.S. defense technology (both products and information) against export to adversary nations. However, because re-exports from a primary recipient are also covered, information shared with close allies (e.g., Five Eyes and NATO) can also be subject to ITAR review and restrictions. This process can impede S&T information sharing. Currently only one nation, Canada, enjoys a limited ITAR exemption.[10] This problem has been recognized for many years.[11] Extending the Canada exemption to selected other allies, as previously suggested,[12,13] could be extremely helpful to IC collaborations. Technical interaction with close allies will be constrained unless exemptions or other means of narrowing ITAR restrictions can be found.

---

[10] Trade Commissioner, Government of Canada, "Export Controls and Canada," updated June 6, 2021, https://www.tradecommissioner.gc.ca/sell2usgov-vendreaugouvusa/procurement-marches/export-cont-export.aspx.

[11] See, for example, U.S. Congress, U.S. House of Representatives, Committee on Science and Technology, *Impacts of U.S. Export Control Policies on Science and Technology Activities and Competitiveness*, 111th Cong., 1st sess. February 25, 2009, https://www.govinfo.gov/content/pkg/CHRG-111hhrg47610/html/CHRG-111hhrg47610.htm.

[12] J.A. Lewis, N.D. Wright, and G. Rees, 2021, *Innovation with Allies: Practical Paths Forward*, Washington, DC: Center for Strategic and International Studies, May 26, https://www.csis.org/analysis/innovation-allies-practical-paths-forward.

[13] T. Bromund and D. Kochis, 2017, *How to Expand Defense Trade Cooperation Between the U.S., the United Kingdom, Australia, and Canada*, Washington, DC: The Heritage Foundation.

Establish IC Standards and Processes to Ensure High-Quality Research, Development, Test, and Evaluation

R&D alone does not create useful products; S&T advances must be transitioned to high technology readiness levels and engineered into functional systems, all of which requires a focus on engineering and engineering expertise. The committee's experience with the IC, and its discussions with its S&T managers, revealed that the IC is uneven in its ability to turn research advances into high-value tools. Missions like the National Reconnaissance Office's mission to build and launch satellites are inherently large scale and large budget, which supports substantial engineering staffs both in-house and at contractor facilities. But other, smaller, programs do not support extensive engineering capacity. The process can be heavily dependent on the skills of contractor organizations. While the IC cannot (and should not) attempt to replicate those skill sets, the IC needs a high level of engineering savvy internally in order to play its own role in technology maturation.

The IC currently lacks a mechanism to develop and enforce engineering standards across the IC agencies, and to provide engineering support to individual programs on an as-needed basis. Many failures or delays result from immature technology being integrated without sufficient testing and maturation, and/or from gross underestimates of funding required to properly test and integrate new technologies. Standards cover aspects such as requirements for independent technical reviews and best practices for testing and evaluation of hardware and software systems at various levels of maturation. Other standards could pertain to topics such as validation and verification methods to be used during design and modeling, standards for configuration management and control, and best approaches for risk assessment and management. When engineering failures do occur, ODNI might oversee a failure review board to determine the root cause of the failure, to promulgate the lessons learned across the IC, and to update standards, practices, and procedures to reflect those lessons learned.

Strengthen S&T Leadership Structure

These Foundational Steps would benefit from the establishment of a new position within ODNI that the committee tentatively calls the chief technology and innovation officer (CTIO). The purview of this position would span the entire technology innovation spectrum, including early-stage R&D (often called "S&T"), experimentation and prototyping, and the system engineering discipline necessary to ensure successful integration of new technologies and capabilities into systems and operations. The latter two items often do not receive sufficient attention when questions are raised about S&T investment and engagement strategies. But ultimately, S&T investments, no matter how large, are of minimal value if standards and best practices are not established and followed for experimentation, prototyping, and system engineering, as these are the activities that translate the potential value of the S&T investments into true operational value.

One key role of the CTIO for ODNI, then, would be to establish and promulgate the expectations, technical standards, and processes that should be followed by all development programs across the IC that are attempting to advance novel technologies and capabilities to a state of operational utility. This responsibility would include:

- Setting the standards for what is an acceptable independent technical review process. These standards would vary depending on the scope and type of program or operation. For those programs of significant funding and/or mission impact, the DNI, at his/her discretion, may request independent technical reviews to be chaired by the CTIO.
- Setting the standards and best practices for test and evaluation.
- Setting the standards and expected processes to be followed by system engineers across the IC, to include design and modeling validation and verification methods, configuration management and control, and risk assessment and management approaches.
- When failures do occur, the DNI may request that the CTIO chair, or select the chair for, a failure review board to determine the root cause of the failure and to publish the lessons learned across the IC. Those the lessons should be included in technical tradecraft training. Standards, practices, and procedures should be updated to reflect those lessons learned.

ODNI, through the D/S&T, provides an essential S&T planning and coordination role. ODNI is charged with coordinating intelligence for the executive branch. The committee believes ODNI is the appropriate locus for coordination of the IC's mission-focused S&T efforts and to provide IC-wide situational awareness of S&T. However, unlike major corporations and some U.S. government departments and agencies, the IC does not have a single individual with authority over the S&T activities of its constituent parts. This idea is not new, and in fact the 2021 CSIS report mentioned earlier recommended that such a position be established: "The DNI should consider further empowering the ODNI Office of Science and Technology and elevating its director to serve as the U.S. IC chief technology officer (CTO)." Additionally, it should be noted that the CIA has announced the creation of a similar position (CTO) in October 2021.

However, it is also important to preserve sufficient agency-level authority with accountability over S&T budget allocations to ensure that mission-related needs continue to be met. An ODNI-level effort should be aimed at supporting mission-related needs and making most efficient use of S&T assets across the entire IC enterprise through enhanced coordination, communication, and opportunities for collaboration.

ODNI's position of Director of Science and Technology (D/S&T) was not established to address this same broad coordination role, but its scope of responsibilities could be expanded. Statutorily, the D/S&T has five responsibilities and authorities:

1. Act as the chief representative of the Director of National Intelligence for S&T;
2. Chair the National Intelligence Science and Technology Committee;
3. Assist the Director in formulating a long-term strategy for scientific advances in the field of intelligence;
4. Assist the Director on the S&T elements of the ODNI budget; and
5. Perform other such duties as may be prescribed by the Director of National Intelligence or specified by law.

Clearly, item 5 allows for the suggested expansion of the D/S&T responsibilities with respect to the IC's S&T. The key needs are that the IC have a clear, high-profile steward of S&T, with authorities and responsibilities that are expansive enough to cover the gaps identified in this chapter. That role must also lead a culture change to elevate the attention paid to S&T across the IC, and adjust practices (e.g., acquisition policies, risk aversion, security trade-offs) as discussed throughout this report. The culture change and selective risk-taking activities needed to achieve this pivot toward stronger S&T capabilities calls for top-down leadership, as is always the case for culture change. To strengthen the IC's capabilities for leveraging S&T, and to generally elevate attention to that critical function, the DNI must first identify someone to lead this effort.

RECOMMENDATION 2.1: The Office of the Director of National Intelligence (ODNI) should consider elevating the priority of science and technology (S&T) by clearly designating an individual to strengthen these Intelligence Community (IC) capabilities. This individual—a chief technology and innovation officer (CTIO)—would report to the Director of National Intelligence, serve as Chief S&T Advisor to the Director, and be charged with the following responsibilities:

- Develop and maintain healthy sharing and participatory relationships across the IC and between it and many relevant domestic and global S&T entities.
- Identify S&T trends with special IC relevance and plan balanced programs of open-source and classified collection and analysis to enable their expedited development and utilization.
- Lead efforts to integrate and coordinate S&T awareness and science and technology intelligence (S&TI). Because S&TI and S&T awareness require different skill sets, and the organizational cultures endemic to each function differ considerably, the CTIO would need to be fully conscious of these differences while fostering shared capacity and understanding to benefit both enterprises.
- Convert this heightened strength in S&T to operational advantage more rapidly and agilely.
- Maintain a diverse, skilled team, selected from within the IC, to be deployed to support the above activities deemed critical to the S&TI mission.

The committee envisions that this CTIO position would enhance coordination of S&T activities among the IC agencies and assist other IC S&T offices to more effectively perform their functions. The CTIO's office would not directly conduct scientific research, compete with the other IC agencies for R&D funding, or diminish the roles and responsibilities of the individual IC S&T managers across the IC agencies; it would not directly control agency budgets and priorities. The addition of qualified technical personnel to support the CTIO's role would likely require increased funding for the CTIO position or for the IC agencies from which the team will be assembled. Recommendation 4.2 and, to a degree, Recommendation 4.3 also pertain to some human resource challenges that call for the CTIO's attention.

The exact roles, responsibilities, and authorities of the CTIO should be developed by the IC's leadership. It is likely that some relevant experience can be gleaned from the CIA's recently launched Transnational and Technology Mission Center, which includes a CTO. ODNI can explore the effectiveness of this CTO position with respect to challenges such as focusing S&T priorities and work to address them. NGA's experience with a CTO might also be illuminating. The primary goal in the committee's vision is for the CTIO to have a very good understanding of what is going on across the IC: what is working well, and what is not; how will external changes affect this; are there new S&T tools that could be used to improve efficiency? Based on a good understanding and process for staying current on what is happening across the IC, the CTIO must understand IC agencies, S&T capabilities elsewhere in the federal government, and domestic and global S&T. The office should have the job of networking across each of these levels for awareness of S&T issues and identify areas that can help solve problems or close gaps internal to the IC. Building on this base, the office would build and execute a process to fulfill these goals and incentivize key stakeholders to contribute to them. The CTIO office would also coordinate S&TI across the IC, coordinate the IC's various S&T efforts, and serve as the S&T advisor to the Director of National Intelligence and the Principal Deputy Director. More generally, the committee suggests the following responsibilities:

- Outward-facing responsibilities
  — Establish and operate a coordination mechanism for the S&TI carried out across the IC. One possible element could be to recruit and assign technically astute and operationally savvy IC S&T liaison officers to work closely with non-IC entities and integrate their findings and insights to maintain an awareness of the global S&T landscape.
  — Establish an S&T evaluation and assessment activity that analyzes the information gleaned from activities described in the previous bullet and raises attention within the elements of the IC that are best suited to utilize the information, and to the DNI; this activity can also feed special requests to the Liaison Officers and open-source analysts.
  — Broaden the IC's engagement with relevant federal laboratories, FFRDCs, and the DoD UARCs, all of which can also provide indirect links to private industry S&T. The CTIO should facilitate partnerships and bilateral arrangements when opportunities to advance IC impact can be achieved via collaboration.
  — Expand where possible the IC's collaborations with unclassified basic research with potential applications to the IC mission and pre-competitive R&D within private corporations and academic institutions.
  — Coordinate international S&T activities across the IC and make international S&T a higher priority, such as through more open engagement with allies.
- Inward-facing responsibilities
  — Serve as a high-level champion for the IC's efforts to develop or acquire, and deploy, game-changing S&T from wherever they may arise. This might potentially include a CTIO funding mechanism, but promising technical base capabilities would still be managed by individual agency program managers for subsequent proof-of-concept, tailoring, and deployment.
  — Establish and promulgate technical standards and processes to lead IC entities in transitioning S&T to reliable, high-quality operational capabilities (e.g., standards for test, evaluation, and validation).
  — Ensure (within the constraints of necessary compartmentalization) that S&T best practices and S&T results from each agency's S&T activities are shared among constituent members of the IC S&T community.
  — Be aware of gaps and shortfalls across all IC missions to help focus S&T efforts and transition newly acquired insight to the correct spots within the IC.

As noted above, the envisioned CTIO position is broader in scope than that of D/S&T as the latter is currently specified. The committee sees the CTIO as coordinating the entire IC TRL stack from science-based discovery and research, up through development and into implementation (which is captured in the "innovation" part of "CTIO"). This coordination is critical because the current IC system is weak at transitioning internal R&D all the way to implementation; the TRL stack needs to be integrated as it is in successful tech companies. In addition, the CTIO should also consider Manufacturing Readiness Levels, recognizing the potential pitfalls in applying technical improvements to broader scale production. Analogously within the context of S&TI, the IC needs to monitor both what adversaries are developing and how they are implementing those advances.

One option could be expansion of the D/S&T position to also serve the CTIO functions described here; an alternative could be to create a new CTIO position, presumably with the D/S&T as a direct report. The committee is not implying any deletion to the current D/S&T responsibilities (items 1-5 in the numbered list preceding Recommendation 2.1), and it is presumed they will continue to be needed.

In spite of this large span of responsibilities, the committee envisions that this CTIO entity would be small, lean, and agile in providing support and coordination across the IC without creating additional bureaucratic hurdles. The committee notes that appropriate staff support for the CTIO would be critical for success in meeting the stated objectives.

The committee does offer one specific recommendation, based on the section above on "Improve Methods for Collecting and Using Open Source Information, with Emphasis on S&T." The 2021 CSIS report *Maintaining the Intelligence Edge: Reimagining and Reinventing Intelligence Through Innovation*[14] reiterated this call:

> IC mission centers must simultaneously move to integrate OSINT into analytic processes and tradecraft. Analysts should view OSINT as a foundational "INT" alongside traditional clandestine intelligence collection in informing and driving analytic judgements. OSINT is the area where application of AI and ML can show early success, largely because OSINT is so vast and so in need of careful curating. A key objective would be to enhance timeliness and relevance of information provided to policymakers and to understand what they may already have absorbed from independent access to open-source data so as not to duplicate reporting. Analysts should focus on integrating what they learn from open sources with other aspects of big data and with secret intelligence to produce the most complete picture of adversary plans and intentions.

The study committee endorses this view and emphasizes the centrality of these recommendations to effective IC use of OSINT. A detailed implementation of how, institutionally, the IC should best increase its intake, analysis, and utilization of OSINT could be the subject of its own study. In such a study, it would be helpful to survey current U.S. government and non-governmental sources of published OSINT, to analyze what is working well, and to determine how the IC could best complement and work with current non-IC efforts. The committee suggests that the recommendations of the CSIS report cited above be considered first.

RECOMMENDATION 2.2: The Intelligence Community should increase its ability to mine open-source science and technology (S&T) information while remaining consistent with prevailing policies and laws regarding privacy protections. It must enable integrating open-source S&T information with classified intelligence.

---

[14] CSIS, 2021, *Maintaining the Intelligence Edge*, p. 15.

3

Leveraging the S&T Activities of Other Federal Agencies

The federal government is a major force in U.S. science and technology (S&T), particularly in funding basic research, applied research, and early technology development. While the U.S. government is no longer the dominant driving force in global S&T that it was in the mid-20th century—when it accounted for nearly half of the world's research and development (R&D) spending[1]—the federal government still supports a significant amount of the global R&D effort. In addition to funding, the federal government is a major performer of R&D, maintaining hundreds of its own laboratories where federal, academic, and private-sector research staff work. Roughly half of this capacity is in the national security sector, of which the Intelligence Community (IC) is an integral part. The government maintains formal mechanisms for R&D planning and coordination across government agencies, including multiagency R&D initiatives; in addition, ad hoc arrangements are available to IC agencies.

Because of the increasing importance of S&T to accomplishment of the IC's missions, and the increasing competition the United States faces in this space (especially from China), the IC needs to take better advantage of all of the S&T capabilities that it can leverage. In that light, it would benefit the IC to do more to leverage the very large S&T footprint that already exists in the U.S. government, which includes large S&T investment budgets;[2] the funding, operation, and staffing of major laboratories; and the maintenance of networks with researchers and S&T experts in academia and industry both in the United States and abroad. The IC does this to some extent, but the committee sees opportunities to do more through mechanisms such as cross-government R&D coordination activities, staff exchanges, and cross-agency R&D.

Strengthening such leveraging could afford the IC opportunities to (1) receive the results of that work by making appropriate arrangements; and (2) find and cooperatively fund specific R&D projects that are of interest to the IC as well as to other agencies. Increased interactions with other agencies also could provide benefits such as staff exchanges and shared assignments. Of course, in pursuing increased interactions, important issues have to be resolved, particularly those associated with the mismatch between IC security needs and culture, and the

---

[1] See, for example, American Association for the Advancement of Science, "Historical Trends in Federal R&D," https://www.aaas.org/programs/r-d-budget-and-policy/historical-trends-federal-rd, accessed August 7, 2022; U.S. Library of Congress, Congressional Research Service, 2022, *Global Research and Development Expenditures: Fact Sheet*, R44283, updated September 14, https://sgp.fas.org/crs/misc/R44283.pdf.

[2] Intelligence Community (IC) research and development (R&D) budgets are generally classified, so this report cannot present a direct comparison of IC science and technology (S&T) to the rest of the U.S. government. However, several experts have commented to the committee that IC R&D funding is a small part of total federal R&D funding.

norms of open research environments. There can also be appearance issues, and risks posed by the fact that many foreign nationals are engaged in the country's open R&D. Also, in some cases it is important not to reveal that the IC is interested in particular areas of research.

In a very general sense, the IC has the same goals as most other organizations that depend on S&T:

- Maintain awareness of what is being researched and developed (everywhere that matters).
- Develop and maintain understanding of the implications and potential of knowledge and developments.
- Gain access (preferably exclusive access, or at least better access than adversaries and competitors) to exploitable results; and move to exploit those results in a timely manner.
- Know and understand what adversaries and competitors are doing.
- Share awareness of new S&T developments with other government agencies close allies on a reciprocal basis.

However, unlike most other government agencies—and private corporations—the IC performs these functions not just to support acquisition of needed technology-based tools for conducting intelligence activities, but also as part of the rest of its mission space. For the IC, these goals could be restated as:

- Conduct S&T intelligence.
- Analyze and report.
- Be positioned to surprise adversaries.
- Help the nation avoid surprise.
- Help keep secure and secret the essential elements of national security-related S&T.

Because the IC and other parts of the government share a need for S&T awareness, and can benefit from improving that capability, it would be mutually beneficial for affected agencies to improve their communication of S&T information. To the extent that the IC is proactive about this, it could also contribute to building bridges and trust that may be beneficial in other ways as well. This chapter explores a range of options to help strengthen the IC's leverage of the whole of government.

MECHANISMS FOR COORDINATION ACROSS FEDERAL AGENCY S&T PROGRAMS

The Office of Science and Technology Policy (OSTP) within the Executive Office of the President is charged with coordinating S&T activities across and among federal agencies. While OSTP does not have control over individual agency budgets, it does have the implicit authority of the White House, ties to the National Economic Council, and a connection to the National Security Council. (OSTP's Deputy Director for National Security also serves as the Coordinator for Technology and National Security at the National Security Council.) Moreover, each year the Director of OSTP and the Director of the Office of Management and Budget jointly issue a memorandum to government agencies outlining priorities that they each should seek to meet in their R&D budget requests. Other coordination activities exist both bilaterally between agencies, departments and offices, and among more than two entities. Some of these involve IC agencies. For example, several IC agencies participate in multi-agency initiatives coordinated at the White House level.

Multiagency R&D initiatives established by statute--such as the Networking and Information Technology Research and Development program, the National Nanotechnology Initiative, the National Quantum Initiative, and the U.S. Global Change Research Program (USGCRP)—are coordinated by the National Science and Technology Council (NSTC) within the White House. These initiatives are not the only examples of excellent opportunities for the IC to increase its level of collaboration with other elements of the U.S. government.

RECOMMENDATION 3.1: The Intelligence Community (IC) should position itself to take better advantage of opportunities afforded by interagency science and technology (S&T) committees and other contacts with non-IC agencies that have substantial S&T activities. More interagency staff exchanges would be helpful, as would more active IC participation in cross-agency research and development. Successful activities on the part of some IC agencies should be studied and mined for best practices.

The Office of the Director of National Intelligence (ODNI) is probably the natural IC office to promulgate more outward-facing involvement by IC agencies because it interfaces with all of them. If a chief technology and innovation officer (CTIO) office is established, as recommended in Chapter 2, it could play the lead in such an effort. To do so, it would need some staff members who are S&T professionals and—we would propose—who do not conduct S&TI (so as not to inadvertently involve intelligence authorities or concerns over intelligence sharing). Those individuals could and should actively engage as equals in interagency S&T activities, with their roles and positions made clear to those they interact with. The ODNI should ensure that the IC has representation on all such multi-agency coordination groups.

PERFORMERS OF FEDERALLY FUNDED RESEARCH AND DEVELOPMENT

The R&D funded by federal agencies is performed at federal laboratories, federally funded research and development centers (FFRDCs), Department of Defense (DoD) university affiliated research centers (UARCs), universities, and private corporations. This section discusses the first three of these mechanisms and identifies approaches that could be helpful to broadening the IC's capabilities in S&T.

"Federal laboratories" is somewhat ambiguous. The committee uses it here to cover (1) facilities staffed primarily by federal employees (e.g., the Naval Research Laboratory; NASA's Goddard Space Flight Center; or the intramural laboratories of the National Institutes of Health [NIH]); and (2) government-owned and contractor-operated laboratories (e.g., most Department of Energy [DOE] laboratories, NASA's Jet Propulsion Laboratory), FFRDCs (42 institutions that are public–private partnerships sponsored by federal agencies[3]), and DoD UARCs, such as the Applied Research Laboratory at The Pennsylvania State University. The extensive and diverse system of federal laboratories range from very specific mission-oriented facilities to large multifunctional laboratories. This broad array of federally funded R&D is many times larger than the IC's capacity for S&T, and it offers extensive opportunities for individual IC agencies and the IC as a whole. The focus here is on R&D supported by DoD and DOE. NIH is another major system of federal R&D, which, like DoD and DOE, supports in-house R&D efforts as well as extramural grants to non-federal researchers. However, to date the IC has not been closely tied to NIH work, even though there are clear overlaps in interests. It is likely that the IC's interests in biological R&D will grow, as referenced in several places in this report, and the ideas presented in this chapter for leveraging DoD- and DOE-funded R&D would apply to NIH-funded R&D as well.

DoD Laboratories

The three military departments maintain among them 20 named Science and Technology Reinvention Laboratories. These are government owned and operated institutions that can take advantage of specific legislated authorities including personnel flexibilities and director discretionary funding flexibility. These DoD laboratories form a diverse group, with work ranging from basic research through technology development, engineering design, and equipment evaluation, support, and eventual disposal. Some have a broad range of work, while others are more narrowly focused. The three major military research laboratories—Air Force Research Laboratory, Army Research Laboratory, and Naval Research Laboratory—have a strong, but not exclusive, focus on basic and applied research.

The DoD laboratories can, and do, work for other government agencies, and maintain partnerships with industry and academia—including through UARCs. For example, the Naval Research Laboratory conducts research not only for the Navy, but also work funded by DOE and NASA. Research partnership agreements that DOD laboratories currently have in place with universities and private corporations include Cooperative Research and Development Agreements (CRADAs), Collaborative Research Agreements/Cooperative Technical Agreements,

---

[3] National Science Foundation, 2022, "Master Government List of Federally Funded R&D Centers," updated February, https://www.nsf.gov/statistics/ffrdclist.

and UARC contracts.[4] Such arrangements can be helpful because, as current and former IC S&T managers told the study committee, the time and complexity of writing research contracts in accordance with the terms of the Federal Acquisition Regulations (FAR) undermines the flexibility and timeliness needed for leading edge research. The existence of other mechanisms such as CRADAs provides S&T managers with helpful options and latitude.

University Affiliated Research Centers

DoD sponsors 14 UARCs, which are described as follows:

> Research organizations within a university or college that are established to provide or maintain essential engineering, research, and/or development capabilities through a long-term, strategic relationship with DoD. Each UARC has areas of expertise that are identified as core competencies that it must provide in support of its mission to support DoD.[5]

UARCs can do work for others (i.e., work sponsored by an agency other than the UARC's principal sponsor), and many do. For example, the Navy's Applied Physics Laboratory at Johns Hopkins University (JHU/APL) does work for the Missile Defense Agency, the Department of Homeland Security (DHS), intelligence agencies, the Defense Advanced Research Projects Agency, and NASA.[6]

Federally Funded R&D Centers

Thirteen federal agencies sponsor a total of 42 FFRDCs. Not all of these conduct scientific research and/or technology development (e.g., some primarily produce studies and analysis), but many are major laboratories. In 2020, 10.5 percent of U.S. government R&D spending ($14.9 billion) was spent at FFRDCs.[7]

FFRDCs are private-sector institutions that have unique relationships with their federal sponsoring agencies. They are administered by universities, other nonprofit organizations, or industrial firms, and they have a special relationship with their sponsoring agency to enable them to serve as independent technical experts. The rules governing FFRDCs are specified in FAR Section 35.017(a)(2). All but one of the 17 DOE National Laboratories are FFRDCs.[8] Others are sponsored by DoD, DHS, the Department of Health and Human Services, the National Science Foundation, and NASA. For most of these, funding and tasking come primarily from their sponsoring agencies, but they also do work for other agencies under separate contracts, and work with or for private-sector entities.

STRENGTHENED INTERACTIONS WITH THE NATIONAL LABORATORY SYSTEM

IC agencies within DoD can—and some do—sponsor FFRDCs and UARCs; however, the cost can be substantial. Other mechanisms are available to IC agencies to make use of these institutions. Individual task contracts are another means for engaging the capabilities of these institutions; IC agencies have used task contracts with DOE national laboratories. The committee believes the IC could benefit from expanded use of, and interaction with, the federal laboratories, including FFRDCs and UARCs. Doing so would provide more ready access to published R&D results, and available capabilities at those entities could be enhanced by the IC funding projects or adding funding to existing projects. Personnel exchanges with in-house laboratories could bring expertise into the IC and get researchers knowledgeable about, and interested in, IC R&D problems.

---

[4] N. Gupta, B.J. Sergi, E.D. Tran, R. Nek, and S.V. Howieson, 2014, *Research Collaborations Between Universities and Department of Defense Laboratories*, IDA Document d-5230, Washington, DC: Institute for Defense Analyses.

[5] Assistant Secretary of Defense for Research and Engineering, 2013, "Engagement Guide: Department of Defense University Affiliated Research Centers (UARCs)," OSD Studies and FFRDC Management Office, Alexandria, VA: Defense Laboratories Office.

[6] Acqnotes. "University-Affiliated Research Center (UARC)," https://acqnotes.com/acqnote/industry/uarc, accessed August 7, 2022.

[7] Congressional Research Service, 2020, *Federally Funded Research and Development Centers (FFDRCs): Background and Issues for Congress*, by Marcy E. Gallo, R44629, Washington, DC.

[8] The National Energy Technology Laboratory (NETL) is not an FFRDC, but is DOE's only government-owned, government-operated laboratory.

One DOE perspective sorts the research environments at its laboratories into two categories:[9]

- An open science environment that is broadly open to the best ideas based on merit, and that has deep ties to the academic and international research communities; and
- A closed science environment that is focused on developing science in disciplines that are not broadly published or are closely tied to national security missions and issues, but which can involve international partners on targeted problems.

The full reality is a spectrum of research environments that range from open to closed, often in different parts of a given laboratory. In addition, conscious links have been set up between the open and closed research environments in which coordination is particularly important, such as the case of the Exascale Computing Project.

The IC already has some strong and valuable interactions with selected DOE laboratories. Primarily these are with DOE's three National Nuclear Security Administration (NNSA) laboratories, which deal with national security issues as well as basic science (Lawrence Livermore National Laboratory, Los Alamos National Laboratory, and Sandia National Laboratories), but connections exist with others in the DOE system, including Pacific Northwest National Laboratory, Idaho National Laboratory, and Oak Ridge National Laboratory. The deep S&T capabilities of the DOE laboratories, coupled with the classified environments available at many of the sites, provide complementary services to those the IC receives from industry and academia. The IC's partnerships with DOE laboratories takes advantage of the latter's well-developed capabilities in areas such as nonproliferation, biosecurity, space technologies, cybersecurity and artificial intelligence, and climate and environmental science.

Some DOE laboratory directors expressed a view to the study committee that the most productive research collaborations are long-term relationships between agencies and individual laboratories and multi-laboratory consortia. In contrast, they characterized much of the IC work at the DOE national security laboratories as based on individual transactions between sponsoring agencies and principal investigators; work is done under contractual arrangements that hand specific research tasks to research groups at the laboratories. This approach can produce specific research advances that meet critical IC needs, but it does not provide a mechanism for the IC research leadership to become deeply familiar with the capabilities of the laboratories or with the emerging research that goes on there, or for the laboratories to develop infrastructure that would be particularly suited for IC problems.

This is not unusual for these laboratories. While there are a few cases in which a non-DOE agency supports specific infrastructure at a DOE laboratory directly for their dedicated use—for example, NIH supports a beamline at a DOE light source for its structural biology research, and DHS supports a simulation center at Sandia—for the most part DOE supports the laboratory infrastructure and facilities that are sufficient to meet its own needs, or that provide a DOE-funded service to the broader scientific community. (Outside agencies that sponsor targeted R&D projects do of course contribute through the indirect cost charges.)

Thus, the DOE laboratories are primarily used by IC agencies as a group of highly capable contractors available to solve specifically defined problems. By following this approach, the IC misses an opportunity to tap the S&T situational awareness that exists within the DOE laboratory system, and it limits the chance for those laboratories to learn about, and contribute to, the broader set of S&T questions that may be of interest to the IC. The IC can access and exploit capabilities that the laboratories have otherwise developed, but has little influence in what future capabilities the laboratories choose to develop.

In contrast, if the IC were to develop longer-term strategic partnerships with some of these laboratories, it could better leverage their S&T capabilities. Enduring strategic partnerships provide stability of both funding and interpersonal relationships that lead to the laboratories developing and maintaining the expertise that best serves IC needs. The model of individually contracted R&D does not allow the laboratory business managers to plan ahead with the IC in mind, as they manage workforce and infrastructure. Examples of successful long-term relationships between DOE laboratories and DoD exist, and a couple exist with IC agencies. Methods to streamline access to the laboratories also can emerge from such trusted relationships.

---

[9] D. Kusenzov, 2014, "The Department of Energy's National Laboratory Complex," paper presented to the Commission to Review the Effectiveness of National Energy Laboratories, Washington DC: Department of Energy, July 18.

There have been sporadic attempts to increase engagement between the IC and the DOE national laboratories, particularly the laboratories of the NNSA. For example, about 10 years ago the Director of National Intelligence and the Secretaries of Energy, Defense, and Homeland Security signed a "Governance Charter for an Interagency Council on the Strategic Capability of DOE National Laboratories as National Security Assets." Among other things, this created a Mission Executive Council (MEC) to implement the agreement. This broad approach, involving several large agencies each with different missions and interest, was not as successful as hoped.[10] A direct bilateral engagement between representatives of the IC and representatives of the DOE laboratory complex would be more likely to reach agreement on shared goals and on what each side needs from the other to be able to work productively together.

Finding ways to get the DOE laboratories more interested in areas of R&D of interest to the IC, and getting the IC more involved in planning research directions and capabilities at those laboratories, would be of real benefit to the IC. For one thing, greater involvement with the DOE laboratories on a day-to-day basis could aid the IC in maintaining S&T situational awareness. The DOE laboratories encompass a broad and deep range of S&T and a high caliber of scientists and engineers. The workforce at these laboratories excel not only in performing cutting-edge R&D, but also in developing and discussing new ideas and research strategies that can lead to dramatic advances. These ideas also benefit from the formal and informal exchanges that DOE laboratory personnel have with much broader S&T communities. The participation of IC S&T personnel in such discussions and planning could be very beneficial to the IC.

Other mechanisms exist for getting deeper insights into the work at the DOE laboratories, such as through more rotational assignments (both to and from the laboratories) and greater involvement of IC personnel in committees that advise or review those laboratories.

The IC could benefit in other ways through increased partnership. For example, it might study how the DOE laboratories attract and retain S&T expertise; both DOE and the IC have shared challenges related to salary constraints, hiring practices, and security.[11] The three NNSA national security laboratories have developed extensive practices to attract some of the nation's brightest scientists, interest and train them in nuclear warhead design and stockpile stewardship, and give them a career balanced between work in this very classified field and in open scientific research. The laboratories use Laboratory Directed R&D funds to support exploration of new concepts, especially by early career researchers, thus making for a more attractive research environment. Analogous approaches might be tailored to help the IC meet S&T workforce goals. Three of the DOE laboratories (Lawrence Berkeley, Argonne, and Oak Ridge) have developed fellowship programs to support developing technology entrepreneurs (e.g., Lawrence Berkeley laboratory's Cyclotron Road, Argonne laboratory's Chain Reaction Innovations, and Oak Ridge's Innovation Crossroads), which is another way to attract talent and stimulate fresh S&T ideas. The Massachusetts Institute of Technology Lincoln Laboratory has recently begun something similar.

Overall, the system of laboratories owned and funded by the U.S. government—including, in particular, the DoD laboratories and the DOE national laboratories—is a major S&T asset available to the IC. The DOE system connects national security laboratories that do classified work to open laboratories and to academic and industrial research teams. While individual IC agencies do have connections to these federal laboratories, the IC is far from taking full advantage of this opportunity.

RECOMMENDATION 3.2: The Intelligence Community (IC) should engage in more active partnering with Department of Energy and Department of Defense laboratories (government, federally funded research and development centers, and university affiliated research center laboratories), to take advantage of their extensive infrastructure and capabilities as well as to employ them as a vehicle for expanding the IC's access

---

[10] These topics are discussed in detail in National Research Council, 2015, *Aligning the Governance Structure of the NNSA Laboratories to Meet 21st Century National Security Challenges*, Washington, DC: The National Academies Press.

[11] Department of Energy laboratories are not staffed by federal employees, but rather work for the contractor for each laboratory. They are nearly all government-owned contractor-operated (GOCO) facilities. By contrast, the IC is staffed by federal employees. But there is a commonality of security rules and related hiring practices, and somewhat less of a commonality regarding salary constraints (salaries may be somewhat higher at GOCOs but still not competitive with much of the private sector).

to academic and industrial research and development activities through overt relationships. The IC should increase its engagement with various laboratory program review activities.

Specifically, the CTIO could be charged with developing strategic relationships with the national laboratories, FFRDCs, and UARCs. This strategic relationship should be designed to leverage long-term goals of both the non-IC entities and the IC, rather than to identify short-term, single-project opportunities. This could begin with the CTIO developing a road map toward creating more working level connections, through staff exchange, mutual education and similar programs. An additional idea could be to hold annual meetings among IC-sponsored principal investigators from different laboratories as a useful tool to strengthen ties among them. Finally, the IC could explore whether the recently established CIA Labs might be expanded to support the entire IC and/or whether it would be advantageous to create a similar organization at the ODNI level to further increase cooperation with other federal laboratories.

4

Leveraging Expertise from the Full U.S. S&T Ecosystem

As pointed out in Chapter 1, the U.S. government is only one piece of the domestic science and technology (S&T) environment. For decades, the academic sector has dominated basic research, but recent years have also seen a resurgence of industry-funded research and development (R&D), mostly in development. Many commercial enterprises now are rooted in cutting-edge S&T, such as Internet communication and commerce, artificial intelligence (AI)-based software, and the growing bio-based economy. There are now many areas in which the private sector leads the government, and in many cases the government's access is no better than that of private consumers and even international adversaries. In order for the Intelligence Community (IC) to take advantage of the deep understanding of S&T that resides in academe and industry, this chapter explores how the IC can leverage the full S&T capabilities of the United States that exist beyond the federal government and its partners.

This chapter explores several issues, including the IC's ability to track and evaluate the implications of progress in multiple, rapidly developing S&T fields. The IC needs to become a more active part of the nation's S&T community to maintain situational awareness, while at the same time protecting its own information, particularly its own technical goals and interests. One partial answer to this dilemma might be for the IC to engage academia and the private sector even more broadly than it now does in work on *unclassified* R&D problems. The goal is to increase IC involvement with the national S&T community in productive efforts, because such involvement (primarily through R&D) creates the channels through which S&T awareness is strengthened and opportunities are created to implement new S&T advances. The relationships, personal and institutional, that develop from such collaborations could eventually lead to closer cooperation between the IC and other S&T sectors, eventually enabling later collaborative, classified development work that integrates knowledge earlier obtained into products of use to the IC. Examples of successes and deficiencies of IC collaboration with academia and the private sector were explored in some detail at the first of two workshops held to support this study (see Appendix A).

The efforts of the Department of Defense (DoD) to expand working relationships with academia and the private sector[1] have been fairly successful in this regard over the past few decades and might serve as a model for IC to engage more in S&T collaborations outside its own confines or, indeed, outside the government itself.

DYNAMIC NATURE OF TODAY'S S&T ECOSYSTEM

Before delving into the broader issues outlined in the above paragraphs, the committee provides an example of the dynamism of today's S&T. Such dynamism poses serious challenges to IC efforts to remain current in rapid developments at the forefront of scientific progress. If not actually actively participating in a field, the IC must remain extremely familiar with latest developments, including organizational developments, at the frontiers of knowledge.

As one example, of the rapid evolution of a field, including of its social aspects, the committee discusses recent changes that affect life sciences research in major ways. These kinds of changes, which to some extent are mirrored in other disciplines as researchers seek better ways of expanding knowledge, inevitably offer new paths of advancement for S&T. The IC's S&T capabilities must be strong enough to recognize the potential opportunities afforded by advances in S&T, infer the ramifications for national security, and facilitate communications among its scientists and engineers. But that challenge is heightened when basic tools and approaches to research are themselves evolving.

Over recent decades, the life sciences have undergone a major transformation from what was largely a descriptive science to what is increasingly an informational and quantitative science. This convergence in the life sciences has been widely discussed.[2] For one thing collaboration is becoming much more the norm. For another, "big data" is having and will continue to have an important role in driving discovery in the life sciences, given the digital nature of biological information, such as genome sequences and the amino acid composition and structure of proteins. How to gather, analyze, and store such data is a pressing need. Convergence between biosciences and information science will blend their cultures, but it will take time because the cultures are quite different. Mismatched expectations and norms can influence interactions negatively, although this disconnect is beginning to fall away.

At the same time, other major changes in the conduct of life sciences research are occurring. For example, the global reach of research teams is becoming more the rule than the exception. Another is that scientific publishing in the life sciences is moving strongly toward posting of pre-print papers prior to journal review, much as has been the case in physics and mathematics.[3]

Meanwhile, life sciences research and technology development are increasingly interwoven. For example, biomanufacturing technology advancement is being heavily supported by federal funding with participation from both industry and academia. In the future, biomanufacturing of nonmedical products such as chemical feedstocks, fuels, and materials is projected to play a very strong role in economic development and countering climate change. In particular, the increasing penetration of biotechnical processes in a wide range of technological and economic applications makes this field of greater significance to the IC that it would have been 10 or 20 years ago. Related to this aspect of convergence, it is becoming increasingly common for scientists and engineers to move back and forth between industry and academia. Not only is industrial experience becoming acceptable, it is also starting to be valued by academia, and increasing percentages of both undergraduate and graduate students are targeting

---

[1] The Basic Research Office of the Department of Defense (DoD) has been active in its outreach to academia, industry, and components of DoD itself, to enhance collaborations outside of DoD, to coordinate research programs and to enhance the science and engineering workforce. International partners are included as well (see the DoD Basic Research Office website at http://basicresearch.defense.gov, accessed July 19, 2021). The Services' research laboratories also have significant research and development (R&D) outreach activities to academia, including fellowships, scholarship programs, and collaborations. Notable among such programs and activities are AFWERX, under which the U.S. Air Force organizes partnerships with innovating elements from the private sector, the 14 university affiliated research centers (UARCs) sponsored by DoD research agencies, and DoD's Joint Artificial Intelligence Center (JAIC), which partners with academia, industry, and foreign allies to develop artificial intelligence with mission impact.

[2] See, for example, National Research Council, 2014, *Convergence: Facilitating Transdisciplinary Integration of Life Sciences, Physical Sciences, Engineering, and Beyond*, Washington DC: The National Academies Press.

[3] Additional observations about ongoing changes in the science and technology enterprise are included in Appendixes A and B, which document two workshops held in the course of this study.

industry jobs instead of remaining in academia. All of these trends affect the way the IC can and should interact with this scientific community.

STEPS FOR LEVERAGING DOMESTIC S&T

Improving situational awareness of S&T developments hinges in large part on intensive and regular interactions between IC experts and S&T experts in academia and industry. When such relationships are established, the IC has greatly improved expertise on call, both for routine updates and also in case of urgent mission requirements. The IC workforce, recognized universally as excellent in key aspects of operational and analytical missions, is limited in budget and personnel regarding S&T expertise needed to monitor all of the technical niches relevant to IC concerns in near-real time.

Of course, a major challenge exists in trying to balance security and free discussion of S&T. There are security-generated barriers to scientific collaborations and other interactions (e.g., conferences, lectures, technical visits, work exchanges) between IC scientists and their counterparts working outside the IC. Security barriers between the most sensitive agencies of the U.S. government and others are obviously necessary and have been vital to the protection of the nation since before the days of the Manhattan Project. But these measures have evolved, and their current manifestation is a proliferation of levels of classification, bureaucracy, extreme delays, and concomitant paperwork that rarely can be challenged, even as these impediments threaten the efficient and effective operations of agencies on which the nation relies to protect itself from serious threats.

This does not mean that IC S&T experts do not go to government S&T meetings, and even occasionally to international S&T conferences. But in the experience of the study committee members, IC experts generally sit on the sidelines, even at government meetings, presumably taking in useful information but rarely interacting with their non-IC colleagues, let alone sharing their own knowledge with their non-IC colleagues. In fact, IC scientists have been actively encouraged to communicate minimally with scientists external to the IC, even with U.S. government scientists, still less with scientists who are outside the government or worse, non-U.S. citizens. At open, professional scientific and technical meetings, which do as a matter of course include foreign citizens, the IC participants frequently have to wear nametags identifying themselves as working for the "U.S. Government," rather than for a particular department or office within the government. This practice quickly becomes a clear revelation to the rest of the world: the bearer of this type of identification is an employee of an intelligence agency, trying to remain "under cover." This approach of passive observation does not encourage engagement between IC S&T experts and their counterparts external to the community.[4]

When IC scientists are so constrained in their activities that they are often not able to attend open conferences or participate in research discussions, the IC is deprived of direct, near-real-time knowledge of new, vital developments in S&T knowledge and discovery that could affect its ability to accomplish its missions.

RECOMMENDATION 4.1: The Intelligence Community (IC) should encourage its technical experts to engage more extensively on a professional level with their peers outside the relatively small IC environment. This would involve attending conferences in their respective fields of expertise, making presentations, and giving talks at other institutions, all with home agency support regarding travel, leave, and expenses. In addition, IC agency experts should be rewarded for inviting outside scientists and engineers to give talks at their home IC agencies. If the proposal to establish a chief technology and innovation officer within the Office of the Director of National Intelligence is accepted, that office would be an ideal place to encourage and oversee these practices.

All of this should be done at an unclassified level, although obvious exceptions could be made in the case of interactions with "outsiders" who possess the required level of clearances to engage on topics at classified levels.

---

[4] A relevant observation from in the commercial sector is found in E. von Hippel, 2005, *Democratizing Innovation*, Cambridge, MA: MIT Press. Chapter 6 of that book which provides a discussion on how the informal sharing of information among technical experts, whose companies may be in fierce competition with each other, results in a positive outcome for all companies.

Such efforts would allow the IC's in-house S&T staff to better discover new advances in fast-developing fields, furnishing the IC with improved situational awareness of current S&T activities and progress.

A related challenge is the time required for an individual to be granted a security clearance—typically 18 months or more at present. Excellent S&T candidates can face lengthy hiring delays, which can cause top candidates to give up on IC careers. This is a problem for national security experts in many fields, but especially in S&T, where the competition for talent is very high. And it is especially acute for the IC because of the general requirement for a TS/SCI level clearance, whereas much other government (notably DoD) work can be done at levels that do not require as lengthy a clearance process.

A third challenge is trust, or rather, the lack of it, between the IC and the non-government sectors. This lack of trust goes in both directions. The IC is wary of scientists and engineers, and, indeed, all non-IC institutions, if they are outside the orbit of U.S. government security review and restrictions. However, outside technical experts have knowledge, experience, and capabilities that could be useful to the IC. This knowledge includes the latest developments in S&T. On the other hand, not only elements of academia, but also some scientists and technologists working for private corporations, tend to regard our national security apparatus warily because of disagreement with some past activities and thus to be avoided.

As noted earlier, a valuable and time-tested mechanism to facilitate communications between the IC technical community and the rest of the country's S&T enterprise is to institute job rotations (in both directions) between the IC technical community and other government agencies, the private sector, or academia. Rotations foster the two-way transmission of knowledge, allowing IC scientists to be better informed of cutting-edge S&T and for non-IC experts to become better informed of the IC's culture, interests, and mission-oriented approaches. Multiple, intricate levels of classification and clearance procedures hamper the efficient rotation of individual experts in either direction—both from the IC to outside positions, and from the broader enterprise into an IC agency. A number of IC agencies are, in fact, proceeding to overcome these obstacles in well-defined cases—one example is the new Central Intelligence Agency (CIA) Technology Fellows Program, announced by Director William Burns in October 2021.

RECOMMENDATION 4.2: To institutionalize increased professional interactions between Intelligence Community (IC) science and technology experts and the rest of the technical world, IC agencies should consider establishing more rotational positions for leading researchers from academia and the private sector, including start-up and venture capital communities. Because SCI level clearances often require a long approval time and impose lifelong prepublication restrictions, some of the rotational positions should be established at both the unclassified and SECRET levels. The lower security level is often subject to a quicker security clearance process than is now practical for the higher level clearances more typical of IC staff. That said, because TS/SCI clearances are usually standard for IC staff, they should be expedited when possible for rotational positions.

As mentioned earlier in this chapter, DoD has built a strong toolkit of mechanisms for R&D partnering. The successes of DoD in broadening its interface with the wider S&T enterprise might provide potential models for the IC going forward.

To help decide on actions by the IC that could improve its leverage of the S&T landscape, it is instructive to examine past exemplars of cooperative R&D activities between the IC and external sectors. This examination may help determine what types of arrangements can be productive. We consider examples of collaboration with academia, federally funded research and development centers (FFRDCs), and industry. In the following paragraphs, the committee presents examples of programs that have engaged a significant number of researchers outside the defense and intelligence communities in work to benefit national security. Mostly originating in DoD or the IC, these efforts have often endured for many years and have resulted in ongoing connections and valuable work between the government's national security community and outside scientists and technologists, which is the measure of success.

An example of a fruitful interaction between the IC and academia is the National Security Agency's (NSA's) longstanding practice of providing grants to researchers at the University of California, Berkeley, for completely unclassified research. (Berkeley does not engage in classified work.) This has produced advantages for both sides.[5] NSA and Berkeley also collaborated through RISELab,[6] which developed SPARK, a state-of-the-art open software for big data analysis. A substantial commercial entity emerged, NSA gained early insights in the technology, and faculty earned recognition for moving technology into engineering practice.[7] Another approach is the university affiliated research center (UARC), such as the Virginia Tech Applied Research Corporation, which brings together academic researchers at Virginia Tech with government sponsors of research, including some from the IC. The recently inaugurated UARC at the University of Maryland, the Applied Research Laboratory for Intelligence and Security, is thus far the only UARC devoted principally to intelligence issues.

Other successful interactions between the IC and academia have been developed by the Intelligence Advanced Research Projects Activity (IARPA), which invests in research programs that tackle real challenges facing the IC. Two of its offices, the Office of Analysis and the Office of Collection, sponsor long-term research across many fields, such as natural language processing, machine translation, biometrics, genomics, machine vision, quantum computing, and power technologies. Results of these programs are published in the open literature and made available through arxiv.org. In addition to broad agency announcements inviting proposals, IARPA runs occasional contests that spur achievement in critical areas, such as open cross-lingual information retrieval or uncrewed aerial vehicle (UAV)-captured imagery.

NSA supports four laboratories that bring together academia and industry to advance science relevant to NSA. Its in-house Laboratory for Advanced Cybersecurity Research conducts mission-focused research to automate network defense, create trusted hardware and software solutions, advance the use of formal methods, and develop tools to enhance cyber threat intelligence. Its work complements work by other NSA in-house researchers by connecting staff with industry, government, academia, and colleagues around the globe. NSA's Laboratory for Physical Sciences combines expertise in the physical sciences and engineering to tackle issues such as quantum and high-performance computing systems or advanced manufacturing and sensing. It carries out a mix of basic and applied research in partnership with the University of Maryland and is situated close to the university. The Laboratory for Telecommunication Sciences, also operating in partnership with the University of Maryland, is another example of collaboration across academia, industry, other government laboratories and IC mission organizations, with a focus on advanced networking, computing, and telecommunications research. Finally, the Laboratory for Analytic Sciences, at North Carolina State University's innovation campus, is a mission-oriented translational research laboratory focused on the development of new analytic technology and analysis tradecraft. It should be noted that these laboratories are able to do classified work.

The National Geospatial-Intelligence Agency (NGA) has an Academic Research Program connecting NGA with academic S&T experts through academic grants to faculty, hiring visiting scientists to perform research on or off site, and supporting undergraduate research in the geospatial sciences at the four military service academies. The academic grants are awarded in three categories, generally for a 2-year base period with up to three 1-year renewals, and are solicited via an Academic Grant Broad Agency Announcement. The three types of grants awarded are for research initiatives, new investigators, and to create collaboration forums in the United States, of course around research areas of interest to NGA.

The study committee believes that IC S&T agencies should more widely involve external partners in performing unclassified research. R&D often does not need to be classified until the technologies or scientific principles are sufficiently developed to be integrated into applications or operations. R&D cooperation in unclassified research has the possibility of building trust between the IC and its partners, enhancing IC access to cutting-edge S&T, and, importantly, building a foundation for future cooperation in classified research. Collaboration in open areas of research would also make such arrangements far more attractive for researchers, who wish to publish their work in the open literature.

Regarding cooperative work between the IC and FFRDCs, there is an extensive history of collaborations between national laboratories and IC agencies, and also, for example, between NSA and its FFRDC, the Center

---

[5] See Randy Katz remarks in Appendix A of this report.
[6] For more information, see RISELab website at https://rise.cs.berkeley.edu/?s=spark, accessed August 7, 2022.
[7] See Randy Katz remarks in Appendix A of this report.

for Communications and Computing of the Institute for Defense Analyses. These collaborations generally work on classified problems.

A Challenge for the IC: Working with Foreign Nationals Within the United States

One of the tremendous strengths of the U.S. university system is that it draws talent from around the world. A large part of the S&T workforce is either foreign-born or the children of immigrants. Immigrants have been awarded 40 of the 104 Nobel Prizes won by Americans in chemistry, medicine, and physics since 2000. Undergraduate students, graduate students, postdoctoral associates, and senior scholars all come to U.S. universities to study and to develop their research careers, and the vast majority of them stay to build their careers, many becoming U.S. citizens. Universities benefit from engagement with international researchers and students in innumerable ways, and specifically in improving both the quality and diversity of its research and research community. This welcoming environment attracts some of the most talented innovators in the world.

The openness of the U.S. educational system to foreign nationals does not just facilitate gaining access to the world's best technical talent. It also permits communication and interaction to take place within the academic community without the need to ascertain or verify citizenship status—which at best would introduce bureaucracy and delay, foreclosing the serendipitous interactions that have been the source of many promising research approaches. Openness also avoids the risk that researchers will provide or withhold information on the basis of characteristics that they may believe to correlate with nationality, such as race, ethnicity, or accent. Discriminating on these bases could violate the law. Even discrimination on the basis of citizenship, if legally allowable for the purpose of participating in government-funded work, would create a two-tiered status among university students and faculty that most universities consider antithetical to the spirit of open communication and collaboration that they seek to foster. Our best universities will not tolerate segmentation of their unclassified research activities into "U.S.-only" and "possible foreign involvement."

Academia does not pursue openness as a gift to foreign countries or their citizens, although they may well benefit from it. Rather, openness is a necessary condition for research excellence. It is not an absolute right, and there may be situations—such as compliance with deemed export control regulations—in which universities or other research institutions will need to restrict participation.[8] However, any such restrictions will impose a cost on research excellence.

Foreign nationals constitute a significant fraction of U.S. academic researchers, as well as collaborators with U.S. researchers on multi-institutional research teams. U.S. S&T benefits from their participation. Some note that U.S. success at attracting the world's best researchers is a major reason why it produces S&T far out of proportion to its population size. Moreover, a large fraction of talented foreign nationals who are educated at U.S. research universities seek to stay in the United States and become contributing U.S. citizens. On the other hand, there is risk associated with foreign access to sensitive R&D in the United States, which has raised concerns in some government agencies. This can put some research universities in a difficult position: their research programs depend on U.S. government funding and on attracting the best talent from the entire world.

When the IC partners with academic researchers, it should generally be able to accommodate the involvement of non-U.S. citizens. Moreover, when the IC partners with the private sector, it should be able to take advantage of the abilities of high-quality scientists and engineers who may not be U.S. citizens.

At the same time, some research projects must be limited to U.S. citizens because they are classified or restricted in some other way, such as through the International Traffic in Arms Regulations. In addition, there have been recent concerns about foreign government interference in research, including instances in which members of foreign military services came to the United States in order to take knowledge and information back to their government. These considerations necessarily impede the IC from taking full advantage of the university system. Several avenues might be pursued by the IC to help it enjoy robust interactions with universities and the private sector without adding to security risks: one would be to encourage the participation of foreign citizens at

---

[8] Fundamental research intended for open publication is typically exempt from export controls, which address the communication of export-controlled information to foreign national independent of their physical location.

universities or in the private sector in research of interest to the IC, if the work unclassified and, in the case of the private domain, is of a speculative, pre-competitive nature.

Building Trust Between the IC and the Academic Community

The IC traditionally, and for good reason, keeps a low public profile. As a result, a large fraction of relevant academics may not be aware of how their expertise intersects with interests of the IC. In addition, many relevant researchers might be reluctant to engage with the IC because of perceived barriers associated with security. The IC could address these issues in a number of ways. The IC could fund unclassified grant programs that are advertised widely to the academic communities of interest. These would be gateways for academics to become more engaged with, and learn about, the sponsoring agency's interests. Researchers on such grants might later join in classified programs. Because the time scales for academic research projects are determined by the career cycles for graduate students and postdocs, it is critical that grants or research contracts last for 3 to 5 years. In addition, it is important that principal investigators and other performers be recognized and brought together in regular meetings to form a standing community, and that funding be sufficient to support such a community with regular additions of new investigators. This is the approach adopted successfully by IARPA, for example.

The IC could also build student scholarship programs analogous to the National Science Foundation's (NSF's) Scholarship for Service program to support individuals who will work in fields of ongoing concern to the IC.[9] That NSF program, run in coordination with the Office of Personnel Management and the Department of Homeland Security, offers scholarships to recruit and train the next generation of cybersecurity professionals. Scholarship recipients must work after graduation for a federal, state, local, or tribal government organization in a position related to cybersecurity for a period equal to the length of the scholarship. Similar programs do exist in the IC, for example funded by the CIA.

A complementary program might be one modeled after NSF's Industry-University Cooperative Research Centers program, which brings academic institutions together with corporations and government agencies to collaborate in long-standing teams on open and pre-competitive research. The students trained in such programs would be natural recruits for both industry and the IC, and the communities built by such programs could be extremely valuable. The IC may not be able to outbid industry in terms of the salaries it could offer such students, but it can still attract those who are motivated to support national security directly, or who want to work on problems and with systems that only the IC can provide.

Another successful DoD model that might be adapted to the IC environment is the early-career grant programs managed by ONR, AFOSR, the Army Research Organization, and the Defense Advanced Research Projects Agency (DARPA). These are extremely competitive programs; they recruit some of the strongest academic researchers at the stage in which they are shaping their research careers. Some number of past grant recipients have gone on to develop long-term research relationships with the sponsoring agency. The programs are designed to recruit researchers who are citizens or permanent residents and whose research is of relevance to the military; because of the focus on early-career researchers, these programs tend to give special emphasis on emerging areas of S&T. While IARPA and NGA already have grant programs analogous to these DoD ones, other IC agencies could create similar early-career programs, focused on S&T areas that are not well enough covered by the existing IC research enterprise. Such programs are needed because it has become increasingly difficult for U.S. citizens to engage in S&T postdoctoral work, owing to increased inducements from industry.

The Defense Enterprise Science Initiative (DESI) is a less developed pilot program by which DoD supports university-industry research collaborations focusing on accelerating the pace at which basic research is brought to the point of impact on defense capabilities.[10] The first round of topics include power beaming, autonomous UAVs,

---

[9] There would need to be an exception carved out if the student cannot obtain a required clearance: the IC agency would not be able to hire the person. Possibly service in an uncleared position in a non-IC agency could be an alternative in such a case.

[10] See Department of Defense, 2018, "DOD Announces Defense Enterprise Science Initiative to Support University-Industry Basic Research," updated January 3, https://www.defense.gov/News/Releases/Release/Article/1407566/dod-announces-defense-enterprise-science-initiative-to-support-university-indus.

soft active composites, metamaterial-based antennas, and remote sensing. If some of these projects are successful, this model would appear to be applicable to accelerating the impact of new technologies of interest to the IC.

Also of note, DARPA runs the Defense Science Study Group through the Institute for Defense Analyses.[11] This program systematically exposes young U.S. academics in science, technology, engineering, and medicine to national security challenges and encourages them to think and apply their insights to these issues. Over the 35 years this program has been in operation, it has involved over 200 technical professionals, and it has contributed significantly to building relationships between DoD and the academic community.

RECOMMENDATION 4.3: The Intelligence Community (IC) should consider emulating some of the Department of Defense's outreach efforts to scientists and engineers in research and development in order to establish trusted collaborations with academia and the private sector. Some of this is being accomplished in the IC, for example, in programs such as IC Scholarships and IC Centers for Academic Excellence. Such efforts should be expanded significantly to develop a trusted community of external researchers. This would be especially useful for engaging researchers without a long history of working with the IC.

Outreach to build academic partnerships via IC-sponsored centers and grant programs can also help academic researchers navigate the significant bureaucratic obstacles (as reported to the committee) that exist when they try to apply for funding from IC agencies. For new investigators, communications with funders are sometimes difficult to establish, which disadvantages them, frequently even to the point of being unaware of IC agency solicitations.

Finally, it could be helpful for the IC to establish a structure that links it more strongly with academic researchers. A possible mechanism would be an academic liaison network for the IC, modeled after the highly successful Special Operations Liaison Network (SLN or, previously, the Special Operations Support Team [SOST]) program, sponsored by SOCOM, which is global and reaches every element and component of the interagency, intergovernmental, multi-national, and commercial sectors. Such a network would be run under the auspices of the chief technology and innovation officer (CTIO), the position recommended in Chapter 2. It could generate tangible benefits to the IC through shared R&D. It might also directly enhance a university's standing and ability to develop its own competitive research portfolios.

The committee envisions that researchers recruited for such a network would be tenured as a network member for a specified period and cleared at the desired level so they can serve a rotational assignment within an IC agency. While at that agency, the network members could work with IC officials, training and mentoring them to better understand S&T issues and trends. In addition, the members could provide situational awareness and analyses of current S&T developments. Such a program would offer those academics an opportunity to experience the intelligence community firsthand, learning about the IC's culture, challenges, and operational constraints, while broadening the IC's contacts among leading S&T researchers.

If designed properly, such a network could enhance the IC's ability to achieve five key goals:

1. Ensure timely tracking of cutting-edge and emerging concepts, technologies, and research activities to help establish fruitful collaborations across diverse academic partners and the IC.
2. Educate both academia and the IC with the aim of establishing and/or improving relationships between both sectors. This would be the result of increased mutual trust and more open communications.
3. Improve the design and implementation of products and processes for both academia and the intelligence community in the course of S&T collaborations; further, efforts should include participation from small business innovation and university-affiliated, private-sector start-ups.
4. Increase academia's national security awareness by informing decision-makers with the goal of improving information exchange between the academic and intelligence community.
5. Expand IC efforts to recruit students and senior scientists and engineers, whose presence could also help provide continuing education to include S&T updates for IC professionals.

---

[11] See the Institute for Defense Analyses website for the Defense Science Study Group at https://dssg.ida.org, accessed November 14, 2022.

Success for such a network requires that it be built with the principles of engagement and embeddedness in mind. Engagement refers to the person-to-person interactions through discourse, collaboration, dissent, and idea-sharing in a trust environment. Because such relationships are generally ad hoc in nature, and personality dependent, a successful network should include an internal mentoring program to socialize the elements of engagement as well as establish standards for them and their dissemination.

Embeddedness refers to the full-time role of being completely immersed in both academia and the IC for a period of time (perhaps 2 years). This might build on the Intergovernmental Personnel Assignment (IPA) process—a mechanism to allow state and local government, academic, or federally funded laboratory employees to work for the government while "holding their place" at their home institutions—on a non- reimbursed or partially reimbursed basis.[12] The IC's Chief Learning Officers Council (CLOC), Intelligence Learning Network, and the IC Training Council (all of which have roles in training IC officers) could enable or assist in the operation of such a liaison network.

The CTIO recommended in Chapter 2 would be the natural office for spearheading these sorts of efforts to strengthen the IC's connections with the broader domestic S&T enterprise.

IC INTERACTIONS WITH S&T INDUSTRY

The IC is far from a stranger to the R&D sector of U.S. industry. For decades, it has contracted with private corporations for technical equipment and other support for its diverse missions. Most prominently since the rapid rise of Earth observation systems in the early 1960s, the IC and other national security entities of the government, as well as civilian research agencies such as NASA, have worked with the aerospace and other industries to advance their missions. Such collaborative work, essential to both national security and many branches of basic scientific research, has led to strong and long-term relationships between many parts of government and the private sector. The IC has been part of this trend; in great part it has worked with its long-term traditional partners, often large corporations with engineering skills that have been nearly unparalleled in the world.

However, the leading role of the IC in some relevant niche areas of S&T is over. Many of the best technologists now work in commercial industry, particularly in the most game-changing areas for the IC such as AI and autonomy. The national security contracting industry is in similar straits. Contractors like Lockheed Martin or Northrup Grumman face increased challenges in recruitment. Moreover, the monopoly that the IC once held in technologies such as earth observation is over; academic researchers and private-sector firms are now producing visual and multi-spectral imagery and radioemitter geolocation of the sort that only IC systems used to provide.[13] Given this change in the landscape, the IC must now acquire the vast majority of its technology from outside vendors, for whom it may be only a small or medium-sized customer.

It is not easy for large bureaucracies of any sort to move as quickly as successful high-tech firms, and unsurprisingly the IC has not always been able to keep abreast of the explosively fast technical progress of the past two or three decades. But the IC has long been aware of this trend and has instituted programs to introduce new technology. In-Q-Tel invests in dozens of new tech start-ups every year, and those technologies are introduced to at least some intelligence officers. In its short existence, the Air Force's technology innovator AFWERX has brought 1,400 companies into DoD as part of its Small Business Innovation Research program, many of which have been introduced to the IC. Dozens, if not hundreds, of other companies and technologies enter the program via other approaches, such as R&D contracts with IARPA or through other transactional authorities.

Ideally, the IC could also go further to support R&D in the private sector and thereby help to bring new externally developed technology into the IC agencies. Support could include funding or the provision of data. A precedent for such actions lies in the joint partnership among NGA, In-Q-Tel, and several commercial companies

---

[12] Another possibility would be a mechanism such as the Department of State's Jefferson Science Fellow program, which is described at National Academies of Sciences, Engineering, and Medicine, "JSF Opportunities at State," https://sites.nationalacademies.org/PGA/Jefferson/PGA_172323, accessed August 7, 2022. Related information on a necessary memorandum of understanding that provides for the home institution to keep paying the fellows' salaries may be found at National Academies of Sciences, Engineering, and Medicine, "Memorandum of Understanding Guidelines," https://sites.nationalacademies.org/PGA/Jefferson/PGA_061588, accessed August 7, 2022.

[13] See, for example, the Hawkeye 360 website at https://www.he360.com, accessed August 7, 2022, and other private-sector Earth observation firms and platforms.

called SpaceNet.[14] More recently, the Biden administration has formed a task force to look into opening up government data for AI development.[15] The IC could explore options for becoming better connected with the equity capital sector, which is now the largest contributor of innovation capital to the U.S. economy. In-Q-Tel plays a role there, but it addresses only a narrow slice of the venture capital market, so more could be done. It may be worthwhile for the IC to examine the way the Foundation for the National Institutes of Health (NIH)[16] helps NIH interact with and influence private industry, and adapt those insights as options by which the IC could do something analogous. Additional thoughts about how collaboration with the private sector abroad is useful for S&T awareness are given in the Chapter 5 section on "Ways for the IC to Enhance Its Access to, and Awareness of, International S&T."

Beyond data, the IC might be able to provide other intellectual property (IP) in a more strategic fashion, possibly under the aegis of a CTIO recommended in Chapter 2. The technology transfer process set up under the Bayh-Dole Act is, in the view of many, less than ideal. Most of the mandated technology transfer offices employ a "catalogue" approach in which they passively await companies to come and ask for IP to license, without any consideration of what technology *should be* commercialized for strategic reasons; these offices then provide little in that transfer process other than a patent license. Few of the most innovative start-ups have the time, resources, or inclination to do such searches and to take big risks on unproven technology. The IC could improve this process by identifying which technologies are strategically important to commercialize, then supporting declassification of certain of those technologies and communicating that to government IP owners. The IC can also help by improving the transfer process, potentially by providing its own inventors and technicians (or those from FFRDCs with which the IC collaborates) to support licensees as they commercialize newly licensed technologies. The IC could also help by covering some of the often significant costs of patenting new technologies for small start-ups. Improving technology transfer would support IC relationships with commercial companies, help grow the technology innovation and industrial base, and support bigger national level strategic needs to compete with, for example, China's own innovation base.

One way that the IC can support technology transition is to adopt a bridge funding approach. This path would involve funds specifically set aside for in-between contracts and directed at preparing technologies for wider-scale use. The funds would be available for evaluated and tested technologies, following the R&D phase, that need to be used in operations at a relatively small scale. A mechanism would exist for increasing the scale of that funding as they go, until they can get to a full-scale program of record.

A different approach to overcoming the technology transition barrier would be to ask Congress for multi-use funds that can be used to pay for whatever activity (research, procurement, or sustainment) is needed to meet program requirements. DoD has received authority from Congress for multi-use funds for several pilot software development programs. This provides much greater agility during development and implementation of new initiatives. The concept is particularly useful for software and AI development, which in commercial industry are almost always pursued through an agile approach. If granted this authority, ODNI could play a significant role in pushing forward approaches like this to better transition technology into operations across the IC.

RECOMMENDATION 4.4: The Intelligence Community (IC) should adopt more forward leaning policies for working with commercial industry to support joint IC–commercial technology development, such as data sharing, as well as acquisition approaches targeted at more effective scaling and implementation of commercial technologies, such as bridge funding. Note that data sharing should go both ways: from the industry partner to the IC agency as well. In both cases, the IC should, by working directly with industry, become part of the technology development process. Additionally, a benefit to the IC of collaboration with the private sector would be increased science and technology awareness. The IC should adopt more active policies for working with commercial industry to support joint IC–commercial technology development, such as data sharing, as well as acquisition approaches such as bridge funding targeted at more effective scaling and implementation of commercial technologies.

---

[14] See the SpaceNet website at https://spacenet.ai, accessed August 7, 2022.
[15] R. Tracey, 2021, "U.S. Launches Task Force to Open Government Data for AI Research," *The Wall Street Journal*, June 10, https://www.wsj.com/articles/u-s-launches-task-force-to-open-government-data-for-ai-research-11623344400.
[16] See the Foundation for the National Institutes of Health website at https://www.fnih.org, accessed August 7, 2022.

5

Leveraging the Global S&T Community

As discussed in previous chapters, the science and technology (S&T) enterprise is global. Research and development (R&D) investment increasingly is distributed worldwide, with approximately 25 percent of global R&D funded by the United States, 20 percent by Europe, and 42 percent by Asia, including China, Japan, South Korea, and Singapore.[1] Scientists are increasingly attending institutions of higher education, conducting scientific research, and pursuing careers in science, technology, engineering, and mathematics (STEM) in foreign countries. In addition, S&T is increasingly collaborative today; scientists frequently conduct research as part of international teams, and many scientists view themselves as part of a global scientific community. The percentage of scientific publications with authors from more than one country is increasing,[2] and such publications have greater impact in field-weighted citation impact analyses.[3]

It is also well known that U.S. companies engage internationally, locating research and production facilities in foreign markets and selling their products worldwide. Supply chains for products may include contributions and components from many different countries. U.S. companies form strategic R&D alliances by working with foreign companies and academia. Many of these partnerships focus on areas of pre-competitive research, participating in communities of discovery to advance S&T that can then be elaborated upon by participants to compete against each other (and third parties) by developing proprietary products and services.

Given the globalization of S&T, the notion of a "domestic S&T base" does not reflect today's reality except for few niche applications.

The United States greatly benefits from engaging in global research in many areas, and many scientific discoveries or technological innovations occur outside the United States. The U.S. S&T enterprise recognizes that it must engage in open global research. The U.S. research community is working to engage with G7 and European Union efforts to preserve research openness, objectivity, accountability, and integrity. Security is an important but challenging issue for this effort—one in which the U.S. Intelligence Community (IC) should be constructively engaged. The IC must do its best to adapt to that reality and leverage the fullness of the global enterprise. One powerful way for the IC to meet the challenge of assessing and accessing global S&T is to engage with those who are developing R&D. Indirect means, by strengthening the IC's capabilities for tapping into existing S&T networks,

---

[1] National Science Foundation (NSF), 2020, "Science and Engineering Indicators," Washington, DC.
[2] NSF, 2020, "Science and Engineering Indicators," Washington, DC.
[3] R.K. Pan, K. Kaski, and S. Fortunato, 2021, "World Citation and Collaboration Networks: Uncovering the Role of Geography in Science," *Scientific Reports* 2(1):902.

are probably most promising. Many foreign academic or government academic institutions, such as defense universities or universities with strong S&T departments, have cultivated valuable working relationships with their own country's intelligence, defense, and government S&T agencies, as well as with their country's industry and other universities. These academic institutions may be amenable to hosting seminars or symposia that bring together scientific experts from both countries to explore areas of mutual interest for future R&D collaboration.

In its *Vision 2030* report, the National Science Board notes the importance of engaging in the international S&T ecosystem—importance heightened by the increasing non-U.S. share of global knowledge production and the rising impact of international collaboration and knowledge- and technology-intensive industries.[4] International S&T cooperation is also essential to address global challenges that the IC has shown concern about and which cannot be tackled by one nation alone—such as climate change and pandemics—and to share the cost of constructing and operating large-scale research facilities.

However, as noted in *National Intelligence Strategy of the United States of America, 2019*,[5] the IC faces a number of constraints that complicate or preclude its ability to directly engage in the international S&T community, particularly outside the foreign intelligence and national security communities:

- The IC has limited R&D resources—with respect to both R&D funding and human capital—yet it has the mission of understanding (and potentially leveraging) global technical advances in a wide variety of areas.
- As noted already in this report, the IC must always be mindful of security, which is especially critical in the case of international engagements.
- The IC has a constrained ability to interact openly with entities, particularly foreign entities, who may be either unwilling to work with it or who may be *too* willing to work with it (e.g., who seek to build relationships for the purpose of distorting information flowing to the IC, or extracting information from it). These constraints pose challenges for widespread and overt international engagement.
- The IC may have difficulty obtaining access to certain international business and technology meetings and conferences where invitations are based upon a willingness of attendees to reciprocate in sharing information.
- The IC has legal and policy restrictions on undisclosed participation in U.S. organizations (Executive Order 12333 and agency implementing policies).
- Lack of dedicated funding for international engagement is another obstacle to increased international S&T collaboration. The IC does not currently allocate funding for international S&T cooperation, including funding for international travel. Foreign travel to meet with international counterparts, researchers, and research facilities may not be perceived as a necessary IC mission expense. Interviews the committee held with IC S&T managers brought out the fact that our FVEY partners do not feel as limited as the U.S. IC with respect to international travel, so the extensive paperwork and limited funding for our IC experts to engage in foreign travel impedes the U.S. IC from benefiting from more effective international collaboration.
- The Arms Export Control Act (AECA) and the International Traffic in Arms Regulations (ITAR) implementing the AECA have hindered Five Eyes cooperation in R&D in the past. AECA and ITAR create a regulatory regime that does not differentiate between allies and adversaries, absent the existence of a bilateral treaty. Even if licenses are ultimately issued for transferring controlled-information exports to close allies, the delay and bureaucratic challenges can be significant problems in state-of-the-art programs.

In its report funded by the National Science Foundation (NSF) on fundamental research security, the JASON advisory group concluded that the benefits of international engagement far outweigh the risks.[6] In academia, international research engagement includes collaborative research projects between academic researchers of different

---

[4] National Science Board, 2020, *Vision 2030*, NSB-2020-15, Washington, DC, p. 24.

[5] Office of the Director of National Intelligence (ODNI), 2019, *National Intelligence Strategy*, Washington, DC, https://www.dni.gov/files/ODNI/documents/National_Intelligence_Strategy_2019.pdf.

[6] JASON, 2019, *Fundamental Research Security*, JSR-19-21, December, https://www.nsf.gov/news/special_reports/jasonsecurity/JSR-19-21FundamentalResearchSecurity_12062019FINAL.pdf.

countries, incoming foreign students and scholars, university/university partnerships, and collaborations between researchers in U.S. universities and researchers working in foreign campuses of U.S. universities.

While it is very important to manage research security risks, it is also important to preserve the open ecosystem for fundamental research that accelerates knowledge and discovery. National Security Decision Directive (NSDD)-189, dated September 21, 1985,[7] established U.S. policy on the transfer of S&T information. NSDD-189 provides that fundamental research should remain unrestricted to the fullest extent possible, and if protection is needed, classification is the appropriate mechanism. The 2019 JASON report on research security recommended that NSF should support the reaffirmation of NSDD 189.[8] Particular risks that the IC faces in engaging globally are discussed below.

As has been pointed out in other reports, this directive is not a favor the United States conveys to the rest of the world, or some blind allegiance to "open science." Neither is it risk-free. But NSDD-189 is necessary, and those risks are worth accepting, in order for the United States to reap the benefits of fundamental research. As explained in a 2005 Center for Strategic and International Studies (CSIS) report, "This Directive does not assert that the open dissemination of unclassified research is without risk. Rather, it says that openness in research is so important to our own security—and to other key national objectives—that it warrants the risk that our adversaries may benefit from scientific openness as well."[9]

Because the IC's culture and processes may inhibit successful sharing and eventual adoption of emergent technology, the chief technology and innovation officer (CTIO) office proposed in Recommendation 2.1 could consider what the Office of the Director of National Intelligence might do to increase the willingness of IC technologists and contracting officers to take more risk and be more aggressive. Possible steps might involve training for liaison engagement, more employment of flexible contract vehicles—such as other transaction authorities, Cooperative Research and Development Agreements, and others—and other steps to make better use of existing flexibilities.

EXISTING U.S. GOVERNMENT ENGAGEMENT IN INTERNATIONAL S&T COOPERATION

The U.S. government engages in international S&T cooperation through government-to-government programs and bilateral S&T agreements. The Department of State's Office of Science and Technology Cooperation manages a portfolio of bilateral agreements that provide overarching frameworks for cooperation in S&T across numerous scientific fields. Numerous federal science agencies such as NSF, the Department of Health and Human Services, the Department of Energy (DOE), and NASA have their own agreements to engage in S&T collaboration in certain fields or to work together to construct and operate large-scale scientific facilities. For example, the Department of Homeland Security's Science and Technology Directorate's International Cooperative Programs Office oversees bilateral S&T agreements to cooperate in research and technology related to homeland security.[10]

The IC could maximize its awareness of and access to international S&T by making better use of knowledge about international S&T activities already available within other federal agencies. As a starting point, the IC would benefit from developing more connections among the IC S&T international managers within the 18 IC organizations and within other government agencies and offices. In addition, it is worth noting that the Central Intelligence Agency (CIA) Chiefs of Stations in overseas embassies have, as a matter of course, useful insights as well as contacts in S&T areas. There needs to be a concerted effort to take advantage of their knowledge in painting a

---

[7] See U.S. National Security Council, 1985, "National Policy on the Transfer of Scientific, Technical and Engineering Information," National Security Decision Directive-189, September 21, https://fas.org/irp/offdocs/nsdd/nsdd-189.htm. The policy was reaffirmed in a November 1, 2001, letter from Secretary of State Condoleezza Rice to Harold Brown, Co-Chair of the Center for Strategic and International Studies, and again on May 24, 2010, in a memorandum from Undersecretary of Defense Ashton Carter to the Secretaries of the Military Departments.

[8] JASON, 2019, *Fundamental Research Security*.

[9] Center for Strategic and International Studies (CSIS), 2005, *Security Controls on Scientific Information and the Conduct of Scientific Research*, Washington, DC, https://csis-website-prod.s3.amazonaws.com/s3fs-public/legacy_files/files/media/csis/pubs/0506_cscans.pdf, p. 2.

[10] Department of Homeland Security, "International Partnerships," https://www.dhs.gov/science-and-technology/st-icpo, accessed August 7, 2022.

picture of S&T activities, both open and secret, in relevant countries. Coordination among these officials, and among other IC staff working on international R&D cooperation are often "stovepiped." IC S&T staff who work with international partners could be drawn together to produce an internal community that shares insights and best practices. The new CTIO position recommended in Chapter 2, or a designee, could establish such a network.

The IC could benefit from the Department of Defense's (DoD's) extensive international S&T cooperation and situational awareness of global S&T developments. The Air Force Office of Scientific Research's (AFOSR's) International Office, the Office of Naval Research's (ONR's) Global program, and Army's Combat Capabilities Development Command actively leverage world-class fundamental research relevant to mission needs. All three operate offices in London, Tokyo, and Santiago to scout out S&T of potential benefit, enable direct interchanges with members of the S&T community across defense, government, academia, and industry, fund researchers abroad, and encourage the establishment of beneficial relationships between scientists and engineers and their foreign counterparts. ONR, AFOSR, and the Defense Advanced Research Projects Agency also provide direct research support to foreign researchers. The Air Force, Army, and Navy international research offices in London moved in 2020 to the Translation and Innovation Hub at Imperial College.[11] The Navy also has offices in Argentina and Singapore.[12]

At one of its information-gathering workshops, the study committee heard from the head of ONR Global, who emphasized the importance of leveraging "smart people across the globe to work together."[13] According to him, the United States needs to focus on building trusted partnerships engaging in global research networks so as to leverage global S&T expertise. In his view, "the secret sauce for building trusted partnerships and networks is to have an enduring presence." This includes building connections and trust by visiting partners' laboratories and meeting with their researchers. ONR also travels internationally, to countries such as China and Russia, to understand the state of fundamental research abroad, and ONR publishes an annual prospectus that is publicly available. While 95 percent of ONR researchers are civilians who do not engage directly with the IC, ONR's military staff keeps IC counterparts informed of relevant information.

In addition to research interactions, ONR assigns domain specialists and regional specialists to various countries to collect knowledge about S&T developments and has created an ONR research network database. ONR, AFOSR, and other DoD entities are collectively exploring multiple network and knowledge management tools to enable sharing of these networks with researchers across DoD.[14] Even if some of these DoD activities are not transferable directly to the IC, the IC could leverage greater benefit from these DoD international efforts, including their networks and situational awareness of global S&T.

RECOMMENDATION 5.1: Within its mining of open-source information in general, the Intelligence Community should increase the collection of open-source information on science and technology advances and early stage companies in foreign nations. The chief technology and innovation officer could coordinate these activities and potentially assign and/or post specialists to cover key regions and countries.

The 2021 CSIS *Intelligence Edge* report[15] includes helpful discussions of the value of open-source intelligence and also suggests ways to expand its use.

---

[11] M. Lachance, 2020, "Tri-Service Partners Join Technology Transfer Ecosystem at Innovative UK University," Air Force Office of Scientific Research, August 6, https://www.afmc.af.mil/News/Article-Display/Article/2243687/international-research-office-embraces-innovation-ecosystem-at-elite-uk-univers.

[12] Office of Naval Research, 2022, "Global Locations," updated March 18, https://www.nre.navy.mil/organization/onr-global/global-locations.

[13] Capt. James Borghardt, ONR Global, presentation, June 9, 2021, in Appendix B.

[14] See Air Force Research Laboratory, "AFRL/RI International Program Office," https://www.afrl.af.mil/About-Us/Fact-Sheets/Fact-Sheet-Display/Article/2332507/afrlri-international-program-office, accessed August 7, 2022. International Agreements are in place for Australia, Canada, Egypt, France, Germany, Greece, Israel, Japan, Korea, the Netherlands, Norway, Portugal, Singapore, Spain, Sweden, and the United Kingdom. See Office of Naval Research, "ONR International Engagement Office," https://www.onr.navy.mil/Science-Technology/ONR-Global/international-engagement, accessed August 7, 2022.

[15] CSIS Technology and Intelligence Task Force, 2021, *Maintaining the Intelligence Edge: Reimagining and Reinventing Intelligence Through Innovation*, Washington, DC: Center for Strategic and International Studies, https://www.csis.org/analysis/maintaining-intelligence-edge-reimagining-and-reinventing-intelligence-through-innovation.

EXISTING COOPERATIVE AGREEMENTS WITH SELECTED ALLIES

The United States also enters into agreements with like-minded countries such as Japan and the United Kingdom to cooperate in specific scientific fields. For example, in December 2019, the United States and Japan signed the Tokyo Statement on Quantum Cooperation to advance innovation and emerging quantum information science and technology.[16] Such agreements include provisions such as promoting specific studies, conducting personnel exchanges, and developing new tools. In September 2020, the United States and the United Kingdom entered into a bilateral agreement to cooperate on R&D of AI technologies.[17] DoD has more than 30 bilateral and multilateral International Cooperative Program Framework memorandums of understanding (MOUs) to provide overarching mechanisms for RDT&E cooperation with allied and friendly nations.[18]

The IC's own relationships with selected foreign intelligence organizations are described in the next section. Here, a sample of some other national-security relationships that the IC might leverage or emulate is presented.

For example, DoD actively engages in North Atlantic Treaty Organization (NATO) S&T activities. NATO's Science and Technology Organization (STO) leverages S&T cooperation among NATO Allies and partners to maintain NATO's military and technological advantage. With a community of more than 6,000 scientists, and a network that draws on the expertise of more than 200,000 people in Allied and partner countries, STO is the world's largest collaborative research forum in the field of defense and security. Its annual work program includes more than 300 projects with a value of 300 million euros that cover a wide range of fields such as autonomous systems, sensors and electronic technology, information systems technology, hypersonic vehicles, and quantum radar.[19]

These facts are cited only to point out the large scale of this cooperation, which the IC may want to leverage more intensely. In fact, Congress, in a 2020 report regarding innovation within the IC,[20] suggested that the IC consider greater involvement in one particular program, NATO's Innovation Hub at Old Dominion University in Norfolk, Virginia.[21] That Hub is an open innovation platform that brings the NATO end user, S&T experts from academia and industry, and a broad community of experts and innovators together to collaborate on design solutions to NATO challenges.

In recognition of the widespread availability of new technologies and the rapid pace at which they are developing, NATO has undertaken several steps to enhance understanding of the role of emerging and disruptive technologies and accelerate NATO's adoption of these technologies. In 2019, NATO's Defense Ministers approved an Emerging and Disruptive Technologies Roadmap to structure NATO's work in these areas.[22] In 2020, NATO's secretary general created a NATO advisory group on emerging and disruptive technologies composed of 12 experts from academia and industry to advise NATO on adoption of new technologies.[23] And recognizing that most new disruptive technologies are being developed by the civil sector, NATO partners with industry though the NATO Industrial Advisory Group comprising more than 5,000 companies, including small- and medium-size businesses.

---

[16] Department of State, 2019, "Tokyo Statement on Quantum Cooperation," December 19, https://www.state.gov/tokyo-statement-on-quantum-cooperation.

[17] Department of State, 2020, *Declaration of the United States of America and the United Kingdom of Great Britain and Northern Ireland on Cooperation in Artificial Intelligence Research and Development: A Shared Vision for Driving Technological Breakthroughs in Artificial Intelligence*, September 25, https://www.state.gov/declaration-of-the-united-states-of-america-and-the-united-kingdom-of-great-britain-and-northern-ireland-on-cooperation-in-artificial-intelligence-research-and-development-a-shared-vision-for-driving.

[18] One example is the Department of State, 2019, "Memorandum of Understanding Among the Department of Defence of Australia, the Department of National Defence of Canada, the New Zealand Defence Force, the Secretary of State for Defense of the United Kingdom of Great Britain and Northern Ireland, and the Department of Defense of the United States of America Concerning Quintilateral Research, Development, Test and Evaluation Projects," December 17.

[19] North Atlantic Treaty Organization (NATO), 2021, "Maintaining NATO's Technological Edge," updated April 13, 2022, https://www.nato.int/cps/en/natohq/news_182871.htm.

[20] U.S. House of Representatives House Permanent Select Committee on Intelligence, 2020, "Rightly Scaled, Carefully Open, Infinitely Agile: Reconfiguring to Win the Innovation Race in the Intelligence Community," Washington, DC, October, p. 31.

[21] The Innovation Hub is part of the NATO Innovation Network, a federation of nodes from NATO and nations leveraging open innovation and agile development.

[22] D.F. Reding and J. Eaton, 2020, *Science & Technology Trends, 2020-2040: Exploring the S&T Edge*, Brussels, Belgium: NATO Science and Technology Organization, https://www.nato.int/nato_static_fl2014/assets/pdf/2020/4/pdf/190422-ST_Tech_Trends_Report_2020-2040.pdf, p. 114.

[23] NATO, 2020, *Secretary General's 2020 Report*, Brussels, Belgium, https://www.nato.int/cps/en/natohq/opinions_182236.htm, p. 63.

Beyond NATO, other multilateral fora for S&T collaboration include the G7 Technology Ministers. An example of one of their multilateral cooperative initiatives is the 2020 launch of the G7 Global Partnership on Artificial Intelligence (GPAI). The GPAI brings together worldwide experts from industry, civil society, academia, and governments to advance research on AI technical topics including trustworthiness and explainability. It will also explore AI workforce development and approaches to spur AI innovation and commercialization.[24] The G7 also adopted a Research Compact at its 2021 Summit in Cornwall, England, to collaborate on research to respond to global challenges, increase the transparency and integrity of research, and facilitate data free flow with trust to drive innovation and advance knowledge.[25]

EXISTING IC S&T COOPERATION WITH FIVE EYES AND ALLIES

The IC currently partners with allies and foreign intelligence and security services to leverage collective capability, data, expertise, and insights, which it considers to be force multipliers.[26] In particular, the IC works closely within the Five Eyes (FVEY) framework. The Five Eyes countries (Canada, Australia, New Zealand, the United Kingdom, and the United States) are parties to the multilateral UK–U.S. Agreement, a treaty for joint cooperation in signals intelligence dating from World War II.[27] FVEY shares signals intelligence, including methods related to signal intelligence operations. FVEY also shares human and geospatial intelligence and engages in signals intelligence cooperation with other allies, including Japan and South Korea. Collaborative arrangements include assignment of foreign liaisons to partner intelligence agencies.

NGA and the National Reconnaissance Office (NRO) represent the United States on the FVEY Allied System for Geospatial Intelligence (ASG). An MOU, known as "Square Dance," establishes a valuable framework for collaborative R&D and test and evaluation (RDT&E) activities among FVEY partners up to the Top Secret/SCI level. The IC primarily engages in RDT&E cooperation with FVEY through Square Dance. Square Dance meets bi-monthly to discuss selected topics, and interested members enter into cooperative project arrangements (PAs) and equipment and material transfer arrangements.[28] Through interviews with senior S&T personnel, the study committee learned that NGA has approximately two dozen cooperative project arrangements in R&D with FVEY members, and that NRO also actively participates in Square Dance activities, including a federated architecture program of small satellites that harden the satellite network. NRO also operates a classified virtual laboratory where technologies can be shared among the FVEY. The IC is also establishing an open laboratory that will allow the FVEY, academia, and industry to participate. Non-U.S. FVEY partners are frequently from foreign defense agencies (as opposed to intelligence agencies) engaging in R&D. FVEY also engages in personnel exchanges under Square Dance: the FVEY assigns employees to serve at foreign partner research laboratories. Assignments typically span 2 years, with the possibility of renewal for a third year.

The IC also engages in international R&D partnerships with FVEY to draw from allied innovation bases supportive of IC needs. For example, CIA has numerous joint projects with Australia and the United Kingdom, some focused on development of next-generation communications equipment. Partnership with Australia enables the United States to leverage Australia's expertise in unmanned aircraft systems, while the partnership with the United Kingdom draws on strong British expertise in material science and clothing fabrication. CIA modeled these international collaborations on NSA's successful working relationship with the FVEY.[29]

---

[24] M. Kratsios, 2020, "Artificial Intelligence Can Serve Democracy" *The Wall Street Journal*, May 27, https://www.wsj.com/articles/artificial-intelligence-can-serve-democracy-11590618319.

[25] United Kingdom Publishing Service, 2021, "G7 Research Compact," https://assets.publishing.service.gov.uk/government/uploads/system/uploads/attachment_data/file/1001133/G7_2021_Research_Compact__PDF__356KB__2_pages_.pdf.

[26] See footnote 5, National Intelligence Strategy, p. 23.

[27] National Security Agency, "UKUSA Agreement Release," https://www.nsa.gov/news-features/declassified-documents/ukusa, accessed August 7, 2022.

[28] Government of Canada, 2018, "Five Eyes Collaborative Environment," http://www.forces.gc.ca/en/business-defence-acquisition-guide-2016/joint-and-other-systems-825.page, accessed August 7, 2022; Australian Government, Department of Defence, 2017, "Science and Technology Portfolio," https://www.dst.defence.gov.au/sites/default/files/publications/documents/DST_Capability_Portfolio_170217_0.pdf, accessed August 7, 2022.

[29] Interviews conducted by National Academies staff with IC S&T managers between March 19 and June 12, 2021.

The United States would benefit from greater engagement with FVEY. For example, because of their relatively scarce resources and limited access to classified intelligence, some of the FVEY partners have developed strong competency in leveraging open-source information, and the IC has the opportunity to learn more about their successful practices. The IC could also benefit from continuing to get FVEY perceptions of threats from common adversaries. At present, international S&T collaboration between FVEY partners is often ad hoc.[30] FVEY collaboration should consider developing a more systematic approach, strategically matching U.S. and FVEY partners' expertise and facilities.

The 2021 CSIS report, *Reinventing and Reimagining Intelligence*, provided several recommendations for the IC to increase its collaboration with FVEY on joint innovation, sharing of algorithms and data, real-time intelligence sharing, and joint development of technologies.[31] These could include the development of talent pools and engagement with commercial entities to tap global expertise and technologies.

WAYS FOR THE IC TO ENHANCE ITS ACCESS TO, AND AWARENESS OF, INTERNATIONAL S&T

There are numerous ways for the IC to enhance its awareness of, and access to, the global S&T landscape. These involve both enhanced direct engagement and indirect engagement. Some options available to the IC include directly participating in research projects with foreign allies and partners, including FVEY and NATO, Federal agencies conducting collaborative international research, foreign universities, and multinational companies. As noted in Chapter 2, several national strategy documents note the force multiplier effect of working with allies and partners.

The committee recognizes that international engagement poses both opportunities and risks for the IC. There is increasing concern about Chinese talent recruitment programs, which encourage scientists based outside China to form affiliations with, or to take positions with, Chinese research institutions, often while concealing their participation.[32] In addition, some countries engage in unlawful or unethical activities in order to gain unauthorized access to U.S. technology and intellectual property. Numerous lines of effort are under way to address these concerns, including the January 2021 issuance of National Security Presidential Memorandum (NSPM)-33, and a January 2021 report from the National Science and Technology Council.[33] These documents prescribe research security and integrity requirements for federal agencies and researchers receiving federal funding, including disclosure of participation in foreign talent recruitment programs; other funding sources; conflicts-of-commitment; and conflicts-of-interest and research security training. In addition, the National Defense Authorization Act (NDAA) for FY 2021 contains provisions requiring federal agencies to implement disclosure requirements for grantees on all sources of research funding support.

In recognition of the benefits of leveraging the S&T expertise of allies, the National Intelligence Strategy encourages the IC to cooperate with international partners and allies.[34] In its report, "Rightly Scaled, Carefully Open, Infinitely Agile: Reconfiguring to Win the Innovation Race in the Intelligence Community," the House Permanent Select Committee on Intelligence, Strategic Technologies and Advance Research Subcommittee[35] recommended that the IC continue to use foreign intelligence relationships as a force multiplier in developing emerging technologies and engage in greater scientific and technological R&D collaboration with the FVEY, NATO, and allies. The study committee believes that, although proposed collaborative activities should be examined closely

---

[30] National Academies staff interviews with IC S&T managers between March 19 and June 12, 2021.

[31] CSIS, 2021, *Maintaining the Intelligence Edge*, p. 32.

[32] U.S. Senate, 2019, *Threats to U.S. Research Enterprise: China's Talent Recruitment Plans, Staff Report*, Senate Committee on Homeland Security and Governmental Affairs, November 19, https://www.hsgac.senate.gov/imo/media/doc/2019-11-18%20PSI%20Staff%20Report%20-%20China's%20Talent%20Recruitment%20Plans.pdf.

[33] U.S. National Science and Technology Council, 2021, *Recommended Practices for Strengthening the Security and Integrity of America's Science and Technology Research Enterprise*, Joint Committee on the Research Environment, Washington, DC: Office of Science and Technology Policy.

[34] See ODNI, 2019, *National Intelligence Strategy*, p. 4.

[35] U.S. House of Representatives House Permanent Select Committee on Intelligence, 2020, "Rightly Scaled, Carefully Open, Infinitely Agile: Reconfiguring to Win the Innovation Race in the Intelligence Community," Washington, DC, October, p. 31.

and individually to make sure that the benefits outweigh the risks, on balance, the IC would likely benefit from enhanced engagements with like-minded allies and partners and the global S&T community.

In some cases, it will be better for the IC to work with its foreign intelligence and security counterparts with whom it may have trusted relationships, and not engage directly outside its international partners. A more public face, however, would enable the IC to increase engagement in open global R&D activities. Many of today's technologies of interest are widely available to individuals, organizations, and countries beyond the United States, creating a more equitable arena for engagement, and less need for secrecy.

In addition, active participation in collaborative R&D projects would enable the IC to participate in "communities of discovery," a paradigm in which scientists from government, industry, non-profits, or academia form open collaborations that share knowledge and coordinate resources to advance scientific discoveries of collective interest.[36] The IC's active participation in research collaboration of this sort would give it a seat at the table, gaining access and insight into scientific discoveries and technologies from the outset, rather than waiting until research or project results are shared publicly with non-participants.

Overall, a more systematic approach to evaluate capability gaps and to evaluate how FVEY, NATO, and other multilateral and bilateral R&D partnerships could fill them, would enable both the United States and our partners to leverage comparative advantages in S&T. Based on its interviews and workshops, the study committee makes the following recommendation for how the IC could increase the benefits it gains from international S&T.

RECOMMENDATION 5.2: The Intelligence Community (IC) should increase its interactions with FVEY (Five Eyes, the intelligence partnership among the United States, Canada, the United Kingdom, Australia, and New Zealand) and other allies through four steps:

1. Create a more systemic approach to cooperation, which could include having the chief technology and innovation officer develop a multi-year allied science and technology cooperation plan.
2. Set aside funding for international cooperative activities (e.g., personnel exchanges, joint research and development).
3. Support travel abroad to deepen foreign partnerships and build trusted relationships.
4. Develop common talent pools and facilitate commercial cooperation opportunities.

More funding for these activities would incentivize and enhance collaboration with FVEY members and other allies. The CTIO office proposed in Recommendation 2.1 should include an international S&T position that reports to the CTIO. The international officer could coordinate, leverage, and integrate international activities across the IC, make international S&T a higher priority, and implement a more systematic approach to cooperation with the FVEY and other allies.

INCREASED USE OF INTERNATIONAL OPEN-SOURCE INFORMATION

Proliferating sensors and big data are generating increasing amounts of open-source information that can be leveraged more by the IC community and integrated with information from classified sources. The enormous quantity of open-source data from numerous sources creates challenges for IC analysts. It has been challenging to take technical advances achieved in unclassified channels and apply them to the same effect with classified data sources. This includes fusing unclassified and classified information into a single intelligence product.

Open-source intelligence and AI tools are also potential areas for increased engagement with the FVEY, NATO and like-minded partners. One indication of the magnitude of the task of analyzing global open-source information on S&T is that China reportedly has devoted some 60,000 to 100,000 analysts to this mission.[37] It is true that

---

[36] IBM Research, 2021, "IBM Science and Technology Outlook 2021," January, https://www.research.ibm.com/downloads/ces_2021/IBM-Research_STO_2021_Whitepaper.pdf, p. 19.

[37] W.C. Hannas and Huey-Meei Chang, 2021, *China's STI Operations: Monitoring Foreign Science and Technology Through Open Sources*, Washington, DC: Georgetown University Center for Security and Emerging Technologies, January, pp. 11, 17.

Chinese needs for open-source S&T information are different from those of the United States and its Five Eyes partners, and this model is not necessarily one that the United States should emulate. But it does indicate the scale of the open-source S&T information that China is interested in understanding.

It would be beneficial to work on open-source reporting on S&T globally as a collection priority across FVEY, NATO and like-minded partners. The IC could benefit from leveraging our allies' expertise in integrating these sources, in addition to their particular areas of expertise and their foreign language capabilities. The IC could also potentially benefit by learning about our allies' experience with application of AI tools and techniques for open-source information analysis, including use of data and algorithms to determine the credibility and reliability of open-source information to identify and assess the impact of foreign disinformation campaigns.

Two IARPA forecasting programs and subsequent work at Georgetown's Center for Security and Emerging Technology (CSET) could also serve as models for leveraging global open-source information to obtain S&T foresight. IARPA conducted the Forecasting and Understanding from Scientific Exposition (FUSE) program between 2011 and 2017. FUSE conducted searches on open science literature in English and Chinese, as well as patent information, for early detection of technology trends and new technical capabilities to avoid technological surprise.[38] IARPA's Forecasting Science and Technology (ForeST) Program, conducted between 2014 and 2016, used crowd sourcing ("Wisdom of the Crowd") to forecast S&T milestones.[39] ForeST established the accuracy of crowd forecasting on well-defined questions.[40] Building on the work of FUSE and ForeST, CSET developed the Foretell Program based on open literature and patent information. Foretell is designed to identify what fundamental S&T innovations are moving quickly that merit attention from policy makers because they are performance changers, have the capability to be disruptive, or represent significant advancements.[41] Initially focused on AI, Foretell has focused to date on identifying which basic science AI-related areas will become emerging technologies, the relevant interest community, and in which areas the United States and allies have an advantage over China.[42]

The quantity, speed, and scope of available global open-source information on new developments and trends in S&T is vast. Obtaining situational awareness of significant and rapidly moving S&T developments and innovations across the globe and across scientific fields will require a concerted effort at a dedicated operations facility, appropriate platforms, relevant analytical and data expertise, and significant resources. As one high-ranking official at NATO Headquarters observed, this increased digital footprint requires transformations in both digital infrastructure and capabilities.[43]

The IC itself may not be the best place to house such an open-source facility, given legal and policy limitations on its ability to collect domestic information and the fact that its deepest expertise is in the collection of classified rather than open-source information. Options include the establishment of a center, institute, or consortium at a non-profit or academic institution to perform this operational function. Another option is to create a center operated by a combination of non-IC and IC agencies. One potential model is the National Counterterrorism Center (NCTC), which produces analysis, maintains the authoritative database, shares information, and conducts strategic operational planning, notwithstanding considerable differences among its member IC and non-IC agencies in their counterterrorism authorities.

RECOMMENDATION 5.3: The Intelligence Community (IC) should work to establish a center (e.g., a non-profit or at a federally funded research and development center or university affiliated research center) operated

---

[38] Dewey Murdick, CSET, presentation at June 9, 2021, workshop, documented in Appendix B. See also Office of the Director of National Intelligence, "Fuse," Intelligence Advanced Research Projects Activity, https://www.iarpa.gov/index.php/research-programs/fuse, accessed August 7, 2022.

[39] Dewey Murdick, CSET, presentation at June 9, 2021, workshop, documented in Appendix B. See also Office of the Director of National Intelligence, "Forest," Intelligence Advanced Research Projects Activity, https://www.iarpa.gov/index.php/research-programs/forest, accessed August 7, 2022.

[40] Ibid.

[41] Ibid.

[42] M. Page, C. Aiken, and D. Murdick, 2020, *Future Indices*, Washington, DC: Center for Security and Emerging Technologies, Georgetown University, October 19, https://cset.georgetown.edu/publication/future-indices.

[43] Jeffrey Reynolds, Operations Research Officer, NATO Headquarters, Supreme Allied Command Transformation, e-mail, and telecom, June 15, 2021.

external to the IC, focused on open-source science and technology (S&T) information collection. This center should take full advantage of collection opportunities, through a presence at international symposia, where potential competitors display their state-of-the-art efforts in mission-critical areas, such as semiconductors, information technology, artificial intelligence, and machine learning, quantum computing/sensing, biotechnology, and other emergent fields of S&T.

The committee also finds merit in the *Intelligence Edge* CSIS report's recommendation that line officers be empowered to directly liaise with IC open-source experts on foreign S&T.[44] Of relevance is the fact that NATO's Allied Command Transformation is currently working on building its capabilities in open-source intelligence. NATO has encountered some of the same challenges as the FVEYs in establishing an international open-source system given the different laws and policies among the NATO partners regarding permissible access and use of open-source information. The IC and NATO could benefit from exchanging experiences and strategies.

Open-source intelligence is well placed to monitor research and innovation produced in the open by academics and in companies which publish and oftentimes promote their breakthroughs. However, much new R&D and innovation still occurs in secret, particularly in authoritarian countries like China and Russia. Given this, the IC must increase its collection of S&TI through clandestine means. The CSIS report also notes this and provides specific recommendations for improvements in doing this, such as increasing integration and support of collection on S&TI issues through new recruiting, training, and office integration. The study committee agrees with this CSIS report recommendation. Beyond this, closer interaction with other U.S. government agencies, as is now being done in collaboration with ONR and AFOSR, could improve timely awareness of adversary trends in S&T that are not part of general and open communication among the global S&T community.

---

[44] CSIS, 2021, *Maintaining the Intelligence Edge*, p. 16.

Appendixes

A

Leveraging the Future Research and Development Ecosystem for the Intelligence Community by the U.S. Research Community: Proceedings of a Workshop—in Brief[1]

This Proceedings of a Workshop—in Brief was published as a standalone document, so its inclusion here is for the convenience of the reader.

On April 19, 2021, the Intelligence Community Studies Board (ICSB) of the National Academies of Sciences, Engineering, and Medicine convened a virtual workshop on behalf of the Office of the Director of National Intelligence. This unclassified workshop was the first of two data-gathering workshops. Panel 1 on Intelligence Needs and U.S. Industry featured panelists Eliahu Niewood, The MITRE Corporation, and Bernard Meyerson, IBM Corporation. Panel 2 on Intelligence Community Needs and U.S. Academia featured John Forte, Virginia Tech Applied Research Corporation, and Randy Katz, University of California (UC), Berkeley. Panel 3 on Intelligence Community Needs and Federal Research and Development (R&D) agencies featured Fleming Crim, National Science Foundation (NSF), and Chris Hassell, Department of Health and Human Services (HHS). This workshop will help to inform a consensus study report investigating how the Intelligence Community (IC) could leverage the evolving R&D ecosystem to meet its future needs and garner the most value from its investments in science and technology (S&T). The consensus study is being carried out by a committee assembled by ICSB. The consensus study committee has also gathered input over the course of several committee meetings from various experts and will also gather more data from a following workshop in June. The consensus study committee vice chair Michael Marletta, C.H. and Annie Li Chair in the Molecular Biology of Diseases, UC Berkeley, invited panelists and participants to share their perspectives on the central question of this workshop: How could the IC better leverage the R&D work of various sectors of the U.S. research community, including industry, academia, and the federal R&D agencies?

INTELLIGENCE COMMUNITY NEEDS AND U.S. INDUSTRY

Eliahu Niewood, vice president, The MITRE Corporation, helped to stand up MITRE's Cross-Cutting Urgent Innovation Cell, a cross-sponsor endeavor that synchronized operational analysis with prototyping and experimentation. This organization helped to identify and close existing gaps with impactful new technology. Prior to joining MITRE, Niewood served as an Intergovernmental Personnel Act (IPA) Mobility Program participant, advising the director of the U.S. Air Force Rapid Capabilities Office (RCO) on technical issues and helping the Department of Defense (DoD) to embrace an innovation initiative. Niewood pointed out that although RCO fields capabilities rapidly and is one of the more innovative organizations within DoD, it is difficult to break down government bureaucracy and foster change on a larger scale.

Niewood guided workshop participants to a publication by Anthony Vinci, former senior intelligence official and chief technology officer and associate director for capabilities, National Geospatial-Intelligence Agency (NGA), who stressed that "intelligence agencies must break down cultural barriers, invest in technology, and dedicate entire offices to artificial intelligence

---

[1] Reprinted from National Academies of Sciences, Engineering, and Medicine, 2022, *Leveraging the Future Research and Development Ecosystem for the Intelligence Community by the U.S. Research Community: Proceedings of a Workshop—in Brief*, Washington, DC: The National Academies Press, https://doi.org/10.17226/26605.

(AI) and automation-driven intelligence."[1] Niewood noted that AI and automation-driven intelligence are only two of the available technologies emerging from industry's R&D ecosystem that are of interest to the IC; others include high-performance computing, biotechnology, nanotechnology, and microelectronics. He also referenced the 2019 National Intelligence Strategy, which asserted that the IC has to "bolster innovation to constantly improve [its] work." Niewood emphasized that "advances in technology are driving evolutionary and revolutionary change across multiple fronts." Niewood underscored that for the IC to become more agile, innovative, and resilient, it would need to dramatically reshape how it operates and leverages this evolving R&D ecosystem. Otherwise, he continued, the IC risks becoming irrelevant, unable to identify trends in a timely fashion: the nation could be left without the necessary critical intelligence to prosper in a world that is increasingly threatened by climate change, pandemics, inequalities, and the growing capabilities of peer and non-peer adversaries. Despite this recognition that change is needed, the situation has not improved—for example, NGA is still considering how to use machine learning and machine vision to change how reams of imagery are processed; the National Reconnaissance Office (NRO) is slow to adopt approaches to enable rapid evolution and take more risk around space systems; and the Federal Bureau of Investigation (FBI) struggles to integrate large volumes of information. Niewood referenced *Maintaining the Intelligence Edge: Reimagining and Reinventing Intelligence Through Innovation*,[2] which suggested that "there is no shortage of opportunities to apply technology across intelligence missions *today*" and that "the primary obstacle to intelligence innovation is not technology, it is culture." He described this as an *innovation problem*: the federal government has created barriers that prevent the IC from effectively partnering with industry to leverage technology.

Niewood noted that these barriers often arise from acquisition processes. Large acquisition efforts (common in the IC and DoD) are focused on writing and reviewing detailed (100+ pages) requirements. These rigid documents are then distributed to a contractor, who has no flexibility to prototype solutions, experiment, or inject new technology. The focus seems to be on modernizing an existing capability, he explained, instead of on analyzing operator needs and applying a new technology to address a critical mission gap. Bernard Meyerson, chief innovation officer emeritus, IBM Corporation, echoed the problems involved with this "gross overspecification of requirements." He wondered why no feedback mechanism exists and why there is no person of authority to override these processes when a better approach arises. Niewood acknowledged that there are very few organizations in the IC and DoD with this level of flexibility in their large acquisition processes. One exception is RCO, which writes requirements only at the paragraph level and has the systems engineering and analysis capabilities in-house to make trade-offs when appropriate. Lilian Alessa, President's Professor and director of the Center for Resilient Communities, University of Idaho, remarked that requirements can enable precision. She wondered if a better approach could be to develop a core requirement alongside adaptive, context-based requirements. Niewood alluded to a similar model, portfolio-based acquisition,[3] which is effective only if the acquisition team has the ability to make trade-off decisions, to understand whether systems will perform appropriately when shifts are made, and to evaluate associated costs. Giving acquisition programs the system engineering and operational analysis capabilities to make smarter decisions would likely result in better products, he stated. Alessa pointed out that portfolio requirements and adaptive requirements are slightly different, but she asked if these more flexible approaches would also require that the user have the appropriate training for and understanding of potential applications of the technology. Niewood suggested that acquisition should be done with the user and the usability of the technology in mind; if users are given a good product, they will learn how to use it. In terms of creating space for adaptation, Meyerson wondered whether it is the user who lacks the right skill set or those who write the requirements who lack the ability to delegate authority. Niewood observed the strictures placed on acquisition departments, which lack both tools and authority. Meyerson and Niewood agreed on the value of dynamic acquisition—that is, meeting the overall goal in whatever way makes the most sense in terms of evolving technologies and programs. Niewood commented that some sectors of NRO do nimble work in

---

[1] A. Vinci, 2020, The coming revolution in intelligence affairs: How artificial intelligence and autonomous systems will transform espionage, *Foreign Affairs*, August 31, 2021.

[2] Center for Strategic and International Studies Technology and Intelligence Task Force, 2021, Maintaining the intelligence edge: Reimagining and reinventing intelligence through innovation, https://csis-website-prod.s3.amazonaws.com/s3fs-public/publication/210113_Intelligence_Edge.pdf, p. ix.

[3] P.J. Modigliani, 2015, Portfolio acquisition: How DoD can leverage the commercial product line model, https://aida.mitre.org/wp-content/uploads/2018/08/Portfolio-Acquisition-NPS-Paper.pdf.

APPENDIX A

prototyping and experimentation, but this practice is not embedded throughout larger acquisition programs. The IC is challenged by a lack of time and resources (and, at times, a concern for vulnerabilities) to leverage new, possibly expensive, emerging technology across the enterprise.

Anthony Vinci, adjunct senior fellow with the Technology and National Security Program at the Center for a New American Security, commented on Niewood's presentation. Vinci highlighted the differences both within the IC and between the IC and DoD in terms of program manager expertise, contracting officer expertise, and ability to work iteratively with industry. He asked about specific models to leverage R&D that could be applied more broadly throughout individual agencies or across the IC. Niewood pointed to Project Maven as a model of success because it identified specific areas of overlap between a critical mission gap and an emerging technology. Moving away from the large, rigid acquisition process, Project Maven succeeded in delivering capabilities to the user in a meaningful way. Even though this is not an example of partnering directly with technology developers, Niewood continued, it still demonstrates successful leverage of available technology. Donald Duncan, senior advisor, Asymmetric Operations Sector, Johns Hopkins University Applied Physics Laboratory, wondered why the lean start-up methodologies[4] to motivate innovation that were being taught at Stanford University were not embraced by the IC or DoD. Although Niewood was not familiar with that particular initiative, he mentioned that there are differences in the national security model and the venture capital model. Although the latter could be leveraged in some cases, innovation is needed around the IC's biggest mission problems, which often cannot be addressed via the venture capital model.

Niewood reiterated that the IC would benefit most from identifying specific quantitative mission challenges[5] and describing them in an unclassified but sufficiently detailed manner to allow the broader industrial community and the academic community to think about solving those problems with emerging technology. This is a better approach from a national security perspective, he continued, than having a massive strategy. A current challenge for the IC is to minimize the impact of the malign influence campaigns from adversaries that increase political divides within the United States. AI and data analytics could be applied, and metrics could be developed to measure the impact of these campaigns. He remarked that involvement from industry to bring this technology to bear to predict, understand, and deter threats would be useful. Meyerson commented on the prevalence of open innovation and wondered how that could affect the IC's involvement with the R&D ecosystem. Although open innovation might not create an *advantage* for the United States, it creates a level playing field and prevents surprises from an adversary. Niewood advocated for breaking down the barriers that previously prevented the United States from taking advantage of opportunities to engage in that level playing field as well as for increasing open dialogue and open innovation among the broader community.

Committee chair Frederick Chang, Bobby B. Lyle Centennial Distinguished Chair in Cyber Security, Southern Methodist University, noted that while diversity in gender, ethnicity, and age are typical goals for organizations, he wondered about diversity in discipline and expertise (e.g., psychology, biology, or chemistry) in the IC. Niewood acknowledged that the IC has made progress in achieving diversity in technical and non-technical backgrounds. He emphasized the value of diverse teams; different viewpoints lead to innovation, and it is not possible for the IC to predict the actions of other countries and cultures if it is not as diverse as the rest of the world. The IC has been less successful in achieving diversity in gender, race, and sexual orientation.

James Gosler, senior fellow, Johns Hopkins University Applied Physics Laboratory, observed that mission owners tend to take their sensitive problems to organizations with whom they have existing relationships (e.g., MITRE, national laboratories, university-affiliated research centers [UARCs], Lockheed Martin, Northrop Grumman), whether or not those relationships are efficient. He inquired as to how the IC would develop relationships with others who could have insights into leading-edge technology but might not realize their relevance to the IC. Niewood suggested starting with problems that can be described at an unclassified level as a way to attract non-traditional industry partners. Once these relationships with smaller companies begin to take shape, when larger or classified problems arise, "interpreters" can break a problem down into components that can be dispersed among appropriate companies. Robert Hale, adjunct senior

---

[4] S. Blank, 2017, The red queen problem: Innovation in the DoD and intelligence community, https://steveblank.com/2017/10/17/the-red-queen-problem-innovation-in-the-dod-and-intelligence-community/.

[5] E.H. Niewood, 2021, Intelligence after next: Mission-based challenges for the intelligence community, https://www.mitre.org/publications/technical-papers/intelligence-after-next-mission-based-challenges-intelligence-community.

fellow, Center for a New American Security, asked if the IC would benefit from special organizations with extra flexibility. Niewood replied that the IC has a few such organizations that reduce bureaucracy, although those arose to address deeply classified problems—the most important, immediate goal for the IC is to focus on its mission problems.

In his talk as a presenter, concluding the first session of the workshop, Meyerson explained that IBM has been driven by innovation for more than 100 years. He portrayed successful innovation as a "creative disruption" that can change the global business landscape. For example, the world's largest taxi company (Uber) has no vehicles, and the largest popular media platform (Facebook) creates no content. He stressed that the IC would benefit from greater awareness of creative disruptors. Manufacturers are working in virtual and emergent environments, which makes it even more difficult for the IC to stay current. He stated that the IC has struggled to make "big bets" on technology, owing to boundaries that prevent adaptation—it is essential to look both forward and backward to avoid technological surprise. Meyerson described System/360, IBM's original mainframe, as an example of a big bet that IBM spent its entire revenue over 1 year to develop. If System/360 had not been successful, IBM would have been destroyed. Bets can be especially risky in emerging fields, but the benefits of the resulting innovations can be substantial. For example, IBM learned early that it could remain competitive only if it pursued open technology, and it placed a big bet on Linux to replace AIX. This is a situation in which a successful program (AIX) was sacrificed for a new program (Linux): the *technical advantages* of Linux outweighed the *profitability* of AIX. He emphasized that the IC faces similar difficult trade-off decisions, although perhaps more in terms of functionality than profitability.

Meyerson indicated that both the Defense Advanced Research Projects Agency (DARPA) and DoD intelligence agencies have helped to enable some of IBM's big bets on innovation. He shared an anecdote about bipolar transistors that began collapsing after several hours of operation, which prompted the belief that there would be no future for—and therefore no need for investments in—the core of silicon technology. This led the IC to invest heavily in gallium arsenide and indium phosphide, with an interest in optical properties and computing capabilities. However, there were foundational issues that prevented either technology from being useful. A Bell Laboratories paper from the 1950s had demonstrated that the same technologies used to make gallium arsenide and indium phosphide could be applied to silicon, if germanium was added to the silicon. Yet, fundamental work with silicon germanium did not begin until the 1980s. IBM determined that using standard commercial technology (e.g., silicon) could provide a better and much less expensive solution than gallium arsenide and indium phosphide. Meyerson noted that people were interested in the possibility of reducing the cost of a major DoD product, but most expected the technology to fail. DoD and DARPA were the only organizations willing to invest in this work, and IBM came to an important realization during one of its experiments: despite 30 years of research on the need to remove an oxide layer from the silicon at 1,000°C in order to make a transistor (a temperature much too high for silicon germanium to remain intact), that oxide was found not to exist. Thus, IBM successfully created silicon germanium, which prompted a drastic increase in transistor performance in orbit at nearly no cost differential (versus the use of basic silicon) and opened up brand new markets for the IC. He explained that the IC essentially built a dual-use program—IBM's interest was in using the technology for WiFi. Previously, WiFi was a commercial product sold almost exclusively to the military; IBM created a partnership with Intersil, claiming more than 99 percent of the WiFi market for nearly 10 years. Meyerson summarized that silicon germanium technology was enabled by a big bet from the intelligence agencies, and IBM solved a problem on which the IC had been spending so much but making little progress. He added that this potential for flexibility and innovation on the part of the IC should still exist today. He championed the value of partnering with the best experts to innovate.

INTELLIGENCE COMMUNITY NEEDS AND U.S. ACADEMIA

Frances Ligler, Ross Lampe Distinguished Professor of Biomedical Engineering, North Carolina State University, introduced the topic of the second session of the workshop: how the interface between academia and the IC has successfully fostered the translation of important R&D products, as well as how it could improve. Randy Katz, vice chancellor for research, UC Berkeley, explained that approximately 60 percent of the $800–900 million in the university's sponsored research comes from the federal government. NSF and the National Institutes of Health (NIH) are the primary sponsors; the more mission-oriented agencies have less of a presence, as the UC System has a policy that it will not undertake classified research.[6] He commented that, prior to 2016, collaboration in science and engineering research and advanced development was embraced as a positive international norm. Over the past few years, issues have emerged around the security of research and foreign engagement of American universities, primarily with nations perceived to be scientific or geopolitical competitors. An additional UC System policy prohibits any projects that restrict participation based on national origin or citizenship. Katz mentioned that approximately 40 percent of UC Berkeley's international graduate student population is from China, and another large cohort is from Iran, all of whom are highly qualified. When national origin restrictions were being promulgated in certain open research projects during the prior administration, he turned away multiple millions of dollars in potential awards that would have limited participation of international students.

Katz described an Australian website that classifies the ways in which universities in China are connected to the Chinese military, a list that could be used to restrict U.S. academic collaboration with higher education institutions in China. Despite the fact that Tsinghua University is known as one of the most prestigious universities in China and one that has collaborated with U.S. institutions for decades, it is listed as a "high-risk" institution. Yet, the workforce from this and other Chinese universities that is drawn to the United States becomes the "intellectual leaders, faculty members, company founders, and engines of translation of best practice from university to industrial activities." He underscored the importance of understanding the dynamics of international collaboration and what that means for the ability to work on projects of relevance to the IC. Owing to UC Berkeley's large population of researchers who may not have U.S. citizenship and the IC's hesitancy to work in open environments, partnerships with the IC sometimes pose challenges.

Katz provided a few examples of UC Berkeley's interactions with the IC writ large. UC Berkeley's RISELab[7] develops state-of-the-art open-source software. The main software artifact that emerged from the project is Spark, which is widely used for big data analyses. The funding that transferred through DARPA for this project originated with the National Security Agency (NSA)—demonstrating that an intelligence agency can work successfully within an open research project. He also described a successful law enforcement agency collaboration (with the FBI) that occurred during the 1990s. The FBI was particularly interested in small-scale microphones that could harvest energy from the environment for wireless communication. This project propelled an understanding of both the science and engineering of low-power, high-integration chipsets. This project spun off dozens of companies and the Berkeley Wireless Research Center,[8] which works closely with federal agencies and industry to advance research. Both are examples of UC Berkeley's ability—with consideration for its commitment to open research, international collaboration, and an international workforce—to pursue projects of importance to federal agencies with intelligence or law enforcement problems. In closing, Katz said that there is no impediment to working with the IC on cutting-edge technological projects in the kinds of models of collaboration used with NSA and the FBI. The impediment has only been the *perception* that the university's open environment cannot be pursued in the context of projects of interest to the IC. He emphasized that UC Berkeley will not change its commitment to an open environment or its collaboration with international researchers; he expressed his hope that the federal government will continue to expand its acceptance of this model.

Meyerson inquired as to how often programs become truly dual-use, with agency engagement as well as the development of a commercial enterprise. Katz replied that Spark is in the hands of Databricks, a company with an $80 billion valuation. In this case, a substantial commercial entity emerged, NSA gained early insights into the technology, and faculty earned recognition for moving technology into engineering practice. He added that many faculty members have start-up companies that are pursuing the commercialization of technologies funded under federal support. Tomas Diaz de la Rubia, vice president for research and partnerships, University of Oklahoma, wondered if there are opportunities for UC Berkeley to work with the IC on an unclassified project, but as it transitions into higher levels of readiness could partner with NSA laboratories for follow-on activities that would lead to classified products. Katz noted that this could be an avenue to explore. However, he pointed out that collaborations with Lawrence Livermore National Laboratory and Los Alamos

---

[6] Unless in an exceptional case, primarily at national laboratories that are part of the university system's portfolio.

[7] See the RISELab website at https://rise.cs.berkeley.edu/.

[8] See the Berkeley Wireless Research Center website at https://bwrc.eecs.berkeley.edu/.

National Laboratory (LANL) are more complicated for UC Berkeley than a collaboration with the Lawrence Berkeley National Laboratory, which is an open science laboratory. He also mentioned Department of Energy projects that have restricted facility access to individuals based on national origin—UC Berkeley would not be able to accept an unclassified project that would prohibit non-U.S. citizens from entering a facility.

Peter Schiffer, Frederick W. Beinecke Professor of Applied Physics, Yale University, questioned whether the IC is successful in attracting UC Berkeley's best graduate students to its workforce. Katz responded that many aspire to pursue academic careers; however, because there are limited jobs in academia, exposure to industry and government career paths would serve graduate students well. Although compensation is much higher in Silicon Valley than in the Central Intelligence Agency, for example, some graduates (with U.S. citizenship) may be attracted to the science and national security missions of the federal agencies. He noted that NSA has thus far done a better job recruiting technical workforce from universities than the IC writ large, which has not successfully communicated its technical and scientific opportunities, but only 1 of Katz's 52 Ph.D. students works at NSA and none work for law enforcement federal agencies. Ligler asked how the IC could better leverage UC System resources. Katz suggested that the IC learn from federal agencies that have worked successfully with the universities on scientific and technical issues. Ligler also wondered whether IPA experiences with the IC could lead to faculty recognition. Although Katz described his own IPA experience with DARPA as transformational, he had little support at the time from academia. He said that the barriers for an IPA participant with the IC are even greater because those faculty cannot share their research. When faculty work in industry, mechanisms exist to protect company intellectual property (IP) while still allowing visibility to the research community; similar rules of engagement would be valuable for faculty taking leave to work in the IC.

Providing another perspective on the interface between academia and the IC, John Forte, chief executive officer and president, Virginia Tech Applied Research Corporation (VT-ARC),[9] explained that VT-ARC was established in 2009 as a 501(c)(3) nonprofit research organization with a mission to deliver tailored analysis, research, and engineering to address problems of national and global importance. He clarified that although VT-ARC is university-affiliated, it is neither a UARC nor a federally funded research and development center (FFRDC). With public, private, academic, and industry partners, VT-ARC leverages VT's $500 million/year multidisciplinary research and innovation enterprise to build highly collaborative ecosystems that advance science, focus research, and enable the integration of cutting-edge capabilities across multiple domains. One important collaborator is the Hume Center for National Security and Technology,[10] the mission of which is to "perform interdisciplinary research and development, provide experiential learning and education, [and] solve challenges in national security by strengthening technology, standards, public policy, and the national security workforce."

Forte summarized the results of a poll that he conducted among a small number of researchers to whom he had access through VT-ARC to better understand (1) how well the IC works with academia, (2) examples of successful and unsuccessful IC–academia projects, (3) how to leverage R&D outside of the IC, and (4) suggestions for improving IC–academia relationships and transitioning and operationalizing concepts and technologies. Poll respondents suggested that relationships between the IC and academia are "mediocre at best." In general, Forte explained, the IC does not work well or broadly with academia. Informing and funding basic and low technology readiness level (TRL) research is often not a starting point of engagement for the IC, which minimizes the success of future capabilities being operationalized. Recruitment is successful only in pockets, he continued, and lasting relationships with recruiters is difficult. It is also challenging to submit concepts to targeted IC organizations and to get research appropriately scoped—an intermediary (e.g., FFRDCs) is often needed. Most requests for proposals require access to high-side networks, and contract and funding mechanisms vary by agency. There is often a greater focus on near-term operational support (months to 1–2 years) and less emphasis on preparing for the future, he noted. Many university IC-focused hubs and centers exist; however, they are typically only connected to the IC in very specific ways, funding levels often do not support what is required to fully escalate TRLs, and cross-center collaboration and associated connections into IC agencies is confusing.

---

[9] See the VT-ARC website at https://vt-arc.org/.

[10] See the Hume Center for National Security and Technology website at https://hume.vt.edu/.

Forte revealed that in 2019, nearly 4,300 degree-granting U.S. colleges and universities had $84 billion in R&D activities, but that opportunity space is not being leveraged by the IC: university-based applied research is highly capable but is utilized less than that of industry counterparts. Sharing additional results from the poll, he commented that the IC and the federal government more broadly could better understand that working with academia is not the same as working with contractors. Relationships with faculty are unique and offer opportunities beyond leveraging research (e.g., externships, connections to industry partners). Long-term relationships can lead to professional development for faculty and capture talent for the IC—the university environment in the United States has a vast and increasing global pull of faculty and students. Freedom of academic research is an essential part of a university ethos, he continued, and there is an important link among research, students, and faculty, in terms of the funding that is injected into the system (i.e., responsibilities to grant degrees to students and to advance S&T).

Forte's poll respondents listed the following R&D project successes: the IC postdoc program; the Intelligence Advanced Research Projects Activity (IARPA) Open Source Indicators program;[11] any project set up to enable dual-use technologies; the Scholarship for Service Program;[12] longer-term programs with direct relationships; the IC Centers for Academic Excellence;[13] and North Carolina State University's Laboratory for Analytic Sciences,[14] University of Maryland's Applied Research Laboratory for Intelligence and Security,[15] NSA–DoD Student Ed programs (e.g., DoDCySP[16]), and the Senior Military College Cyber Institute. Forte added that when the IC is interested in both project and student outcomes, this often leads to a stronger future workforce, better relationships, and more impactful capabilities. Regarding R&D challenges, poll respondents observed that some projects, particularly those of longer duration, failed to aid in pushing student clearances. Even those universities with hubs and centers that are deep in staff with clearances still have difficulties engaging with the IC. Furthermore, requirements are rarely articulated in a way that ensures that research and products will be aligned with a mission need. Forte emphasized that timelines with universities are different; fielding a capability in 1 year or less under current scenarios is difficult at best, and burying a university under a prime contractor often limits impact and partnerships. Poll respondents provided suggestions about leveraging R&D from other sectors: replicate DoD's approach to a more open R&D environment; search for ways to enable R&D programs that have dual use and to scan industry for S&T that has dual use; and identify counterparts in other agencies with similar goals, allowing them to be the more transparent partner in the relationship with those in other sectors. Last, Forte shared poll respondents' overarching suggestions on improving collaboration between academia and the IC.

- The IC should fund more R&D tasks that require more than months to accomplish and have applications broader than what are now the most common assignments—urgent responses to specific, short-term operational IC needs;
- Devise more transparent time frames for returns on investment by the IC from academic researchers with whom it contracts;
- Enable efforts to demystify the IC and educate academia;
- Determine how to overcome the challenge of non-U.S. researchers at an institution that performs classified research and protection of research from dissemination that is not desired by the IC funder;
- Acquire tools to identify where great S&T is occurring (beyond the "big five" universities);
- Increase long-term funding (money in the academic community facilitates support of graduate students, postdocs, faculty summer salaries, and operation of research laboratories);
- Create and articulate transition plans and partnerships before funding, which may involve different university, industry, and government players, as well as different phases;
- Adopt mechanisms that enable flexibility in research and impact, including indefinite delivery–indefinite quantity contracts, cooperative R&D agreements, educational

---

[11] See the Open Source Indicators program website at https://www.iarpa.gov/index.php/research-programs/osi.

[12] See the Scholarship for Service Program website at https://www.sfs.opm.gov/default.aspx.

[13] See the Centers for Academic Excellence website at https://www.dni.gov/index.php/iccae.

[14] See the Laboratory for Analytic Sciences website at https://ncsu-las.org/.

[15] See the Applied Research Laboratory for Intelligence and Security website at https://www.arlis.umd.edu/.

[16] See the DoDCySP website at https://public.cyber.mil/cysp/.

partner agreements, and partnership intermediary agreements;

- Avoid pitfalls in promising new programs;
- Accept more risk, adopt an ethos of allowing research to fail (fast), and be bold;
- View university R&D providers as partners, not suppliers;
- Determine how to build incentives for publication and commercialization;
- Develop more pathways to aid in describing the research desired by the IC to potential academic participants, including easier access to submit concepts and provide feedback; and
- Gather insights on what the IC needs from a particular project, which will result in more impactful capabilities (i.e., the greater the focus on requirements, the greater the chance of delivering something that is already old upon delivery).

Alessa wondered whether a stigma exists among faculty about those working *for* the IC as opposed to working *with* the IC in the spirit of collaboration. Forte suggested that the IC codify better processes and relationships. He stated that every academic organization is different; it is important for the IC to understand these cultures and avoid trying to change them. Ligler asked about the most effective way to facilitate dual-use applications for the IC. Forte advised looking at the requirements of other government agencies and augmenting those with a series of threshold requirements for research products. Trusted and close relationships between government agencies (e.g., parts of the IC and DoD)—for example, DoD mechanisms for dealing with funding of applied research and the needs for secrecy involving some products—could enable the IC to resolve such dual-use issues. In other cases, it may be possible to reach out directly to specific industry players and pull in their technology, although that requires a specific function for adapting the technology to the mission capability. Ligler added that companies can often surpass the "valley of death" if their product is useful for both civilians and the IC. Forte indicated that the role of intermediaries could be explored in greater depth to build this type of ecosystem. Gosler asked what the IC system could do to help universities better understand counterintelligence issues, thus building a bridge between the IC and academia for the benefit of the nation. Katz commented that universities have to follow existing rules and regulations and emphasized that universities are ready to work with the counterintelligence community on its investigations. However, he asserted that universities are not police forces; they depend on partnership with federal agencies such as the Department of State and the U.S. Citizenship and Immigration Services to help ensure that the people in the campus workforce are permitted to be in the United States. Universities are trained only to be aware of extraordinary occurrences that could be security risks. Forte added that he is sensitive to these issues, but he agreed with Katz that a university cannot restrict freedom of academic research. If universities act like police forces, they will lose valuable research faculty and research dollars; a different mechanism to expose risks in sensitive work would be useful—perhaps via the IC or intermediaries.

INTELLIGENCE COMMUNITY NEEDS AND FEDERAL R&D AGENCIES

Opening the final session of the workshop, Gerald Epstein, distinguished research fellow, Center for the Study of Weapons of Mass Destruction, National Defense University, mentioned the role of government funders to link academia and the IC but also highlighted the government's function as a significant performer of research. F. Fleming Crim, chief operating officer, NSF, provided an overview of the structure of NSF as well as how the organization connects to the IC. The mission of NSF is to "promote the progress of science; to advance the national health, prosperity, and welfare; to secure the national defense; and for other purposes." He explained that NSF supports academic basic research, providing 27 percent of the total federal support for all science and engineering fields through seven directorates—Biological Sciences; Engineering; Mathematical and Physical Sciences; Computer and Information Science and Engineering; Geosciences and Polar Programs; Education and Human Resources; and Social, Behavioral, and Economic Sciences—and two important units—the Office of Integrative Activities and the Office of International Science and Engineering. This year, he continued, NSF is an $8.5 billion agency, and this funding reaches more than 300,000 people. Crim described three modes of NSF connection to the IC: (1) program collaborations; (2) interagency committees; and (3) international issues, particularly in relation to science and security. One example of a program collaboration is that between the NSF Division of Mathematical Sciences (DMS) and NGA, which is funding "Algorithms for Threat Detection" to develop the next generation of mathematical and statistical algorithms to deal with large spatiotemporal data sets and to better understand artificial neural networks. In this case, DMS and NGA jointly

APPENDIX A

prepared solicitations in areas of mutual interest, and NSF reviewed the proposals on the basis of intellectual merit and broader impact. He stated that Industry–University Cooperative Research Centers (IUCRCs) provide great potential for NSF and IC partnerships. NSF provides funding to stand up and administer the IUCRCs, and the participating academic institutions, industries, and government agencies fund and determine the direction of the research. There are nearly 80 IUCRCs, three of which have significant participation from NSA. The University of Pittsburgh is the lead academic institution for the Space, High-Performance, and Resilience Computing IUCRC,[17] which has more than 30 participants, including government organizations, academic laboratories, and industrial laboratories. In 2020, this IUCRC funded 11 projects. The Accelerated Real-Time Analytics IUCRC,[18] led by North Carolina State University, focuses on wrangling rapidly changing real-time data. The Cloud and Autonomic Computing[19] IUCRC is led jointly by Texas Tech and the University of Arizona, with six projects in 2020. Another way in which the IC organizations connect to NSF is by providing supplementary funding for NSF projects (e.g., on trustworthy deep neural networks and on research on cyber-physical systems). He emphasized that all of the aforementioned research is unclassified.

Crim remarked that NSF is also connected to the IC via interagency collaborations. For example, the Climate Security Advisory Council is an interagency committee of 20 agencies, including members of the IC, DoD, and the federal science agencies. The purpose of this council is to anticipate climate change effects on national security interests and inform senior policy makers about risks and opportunities. He also mentioned several Office of Science and Technology Policy (OSTP) and National Science and Technology Council (NSTC) groups that have IC participation, such as National Security Council (NSC)–OSTP, Bioeconomy; NSC–OSTP, Critical and Emerging Technologies; NSTC Subcommittees on Machine Learning and Artificial Intelligence, Future Advanced Computing Ecosystem, Quantum Information Science, and Networking and Information Technology R&D; and an Interagency Working Group on Special Cyber Operations Research and Engineering. Other collaborations focus on international issues related to science and security. For example, NSF programs such as the Rubin Observatory, the Antarctic and Arctic Program, and the Seismology Networks do work that connects to the concerns of the IC, as do the NSTC Subcommittee on Research Security; the Interagency Working Group on Research Security and Training; and various outreach events, such as university–FBI meetings.

Responding to Crim's overview of NSF, Chris Hassell, acting principal deputy assistant secretary and senior science advisor, Office of the Assistant Secretary for Preparedness and Response (ASPR), HHS, explained that when the FBI wanted to explore the foundations of forensic science, it partnered with NSF to initiate programs. Hassell also agreed that IUCRCs provide enormous value for collaborating, moving research forward, and recruiting workforce.

Offering another federal government perspective, Hassell indicated that ASPR has two major areas of focus: (1) deployable medical teams that respond to public health crises in the nation; and (2) medical countermeasure development, via the Biomedical Advanced Research and Development Authority (BARDA). BARDA worked with DoD on Operation Warp Speed during the COVID-19 pandemic to distribute vaccines, therapeutics, and diagnostics. BARDA also works closely with NIH in a systems engineering role to scale up to commercialization and engage with the Strategic National Stockpile. The U.S. Food and Drug Administration and the Centers for Disease Control and Prevention are other important partners for ASPR. Hassell noted that multiple federal agencies would collaborate in the event of a release of a biological agent in the United States. It might first be perceived as a public health situation, but ASPR would coordinate with the IC and law enforcement to better understand whether the release was actually a deliberate attack and address the situation. He emphasized that ASPR, and HHS more broadly, relies on the IC for accurate and timely information to understand the complete landscape of threat determination and analyses.

Hassell remarked that agency collaborators consider what investments should be made to produce more effective medical countermeasures—for example, work on the mRNA platform for

---

[17] See the Space, High-Performance, and Resilience Computing website at https://iucrc.org/node/7661.

[18] See the Accelerated Real-Time Analytics website at https://iucrc.nsf.gov/centers/center-for-accelerated-real-time-analytics.

[19] See the Cloud and Autonomic Computing website at https://iucrc.nsf.gov/centers/cloud-and-autonomic-computing.

the Pfizer and Moderna vaccines began at DARPA and was acquired by NIH, enabling a swift start for COVID-19 vaccine development. It is crucial to understand what countermeasures need to be readily available for the future, whether the disease agent is naturally occurring, modified, or intentionally produced. He stated that for intentional threats—especially those created by genomic engineering, synthetic biology, or advanced biotechnology—the best defenses are the technologies themselves. HHS works with its IC colleagues to better understand and evaluate those technologies to prepare for rapid response. He asserted that data science, automation, additive manufacturing, and nanotechnology are other important technologies that could be brought to bear by adversaries and thus should be understood and applied in the United States. He highlighted an IARPA program, Finding Engineering-Linked Indicators (FELIX), that links advanced genomic sequencing technology with data analytical technology for rapid analysis to determine whether an organism has been intentionally modified for increased potency. Researchers discovered that the same technology could be useful for disease surveillance situations, and HHS is considering how this construct could be leveraged for the public health mission. Hassell emphasized that the FELIX performers came from academia and industry, with funding from the IC.

Epstein observed significant technical overlap and great potential for common purpose between the IC mission and the research being pursued by the science and federal mission agencies. Although there are clear differences in the way security-focused agencies operate (e.g., with a focus on information control and counterintelligence) as compared to science agencies, both types of agencies have widespread international engagement. He wondered if companies in other countries view engagement with the IC more cautiously than that with other U.S. government agencies and asked whether overt engagement with the IC has the potential to interfere with international engagement activities. Crim responded that NSF's conversations with NSA are no more difficult to carry forward than those with NIH, for example, because missions and restrictions still have to be aligned, and all NSF research is unclassified and transparent. This approach helps to ease the concerns of international partners. He added that the burden of being perceived to be working with the IC is actually greater in the domestic academic landscape than in the international landscape. Hassell explained that during his tenure with the FBI, when the organization engaged with the academic community, it assumed that institutions would have preconceived notions about dealing with "the men in black," but that was either not the case or the misconception was quickly dispelled. During FBI-sponsored workshops, several institutions admitted that they did not know what to do when they suspected inappropriate activity in their laboratories. The FBI was able to provide guidelines about who to call and when. The FBI's high level of engagement reinforced the notion that universities and law enforcement agencies are not segregated. In cases in which academic institutions are hesitant to work with a large agency, Hassell noted that it is important to recognize other opportunities—for example, researchers can be engaged with LANL's mission without having to work on the nuclear weapons program. Hassell and Crim championed transparency and improved conversations between academia and federal agencies to better understand one another and to reduce barriers to collaboration.

Chang alluded to editorials opining on whether the United States has declined in science and engineering. He wondered whether NSF systematically monitors the global S&T landscape to understand if, in fact, the United States is falling behind. He asked about the best way to measure this, and whether any action would be taken by NSF in response. Crim replied that NSF houses the National Center for Science and Engineering Indicators, which monitors this issue by collecting and analyzing survey data. Each year, in coordination with the National Science Board, these data are published online. Crim noted that the metrics vary. For example, 10 countries currently dedicate a larger fraction of their gross domestic product (GDP) to R&D than the United States; two decades ago, only one country dedicated a larger fraction than the United States. In terms of the fraction of the GDP that is allocated for basic research, the U.S. federal government only funds one-third now instead of the two-thirds of two decades ago. Crim noted that the United States now competes in a global scientific environment. Chang asked what worries the agencies most about this issue. Crim remarked that if the United States is not vigorous in doing science and creating scientists for the future, it will not be healthy as a nation. He fears the United States failing to exploit the intellectual drive that makes it successful, not only in terms of international competition but also in terms of national quality of life. Chang questioned how NSF would respond if a U.S. adversary started to double its initiatives in a particular field. Given that NSF is likely aware of the technologies for which other countries are increasing funding, Gosler wondered if it would share that insight with the IC, which could then use its capabilities to reach ground truth and help the United

States to determine its next investment. Crim said that NSF holds workshops about emerging technologies to better understand scientific opportunities, but it does not systematically take the steps that Gosler mentioned. Crim explained that there are proposals for legislation to direct NSF to compete in certain areas. But, currently, NSF funds one out of every four proposals that it receives, which means that a few billion dollars for highly rated research is being left on the table. He commented that many researchers believe that NSF's awards are too small or too short in duration, and that NSF is not providing enough of them. Yet even small increases in any of these areas would require a doubling of the entire NSF budget. He reiterated that NSF receives a signal about emerging areas of competition based on the proposals it receives, and it makes decisions about which proposals to fund based on intellectual merit and broader impacts. NSF tries to move where the intellectual opportunities exist, while maintaining the health of its entire scientific enterprise.

Schiffer pointed out that highly skilled people in demand can make a significant amount of money in the corporate sector, without the constraints typical of government work. He asked how HHS recruits the best workforce to identify emerging threats and how NSF encourages people to pursue government service. Hassell emphasized how important it is to get students interested in science at the middle and high school levels. HHS has science, technology, engineering, and mathematics workshops for students, and it provides education on the variety of career trajectories available for federal employees (e.g., FBI agents who are chemists). HHS actively recruits not only from universities but also from pharmaceutical companies and corporations. For purposes of recruitment and retention, HHS tries to create opportunities similar to those in the private sector (e.g., sabbaticals, industrial interactions). Crim agreed with Hassell that concerted efforts are needed to underline the service opportunities of government work. He discouraged recruiting solely from the traditional demographic of science and encouraged increasing broader participation. The other key, Crim continued, is to create a workforce at all levels—that is, scientists with Ph.D. and B.S. degrees as well as skilled technical workers.

Marletta asked what NSF could do better to engage the IC and its mission challenges with the academic and research community in fundable projects. Crim responded that the IC's provision of supplementary funding is an effective model. A memorandum of understanding (MOU) with a joint solicitation is also effective because it focuses on areas of importance to both organizations. MOUs can be problematic, however, in that each one is customized. He suspected that cumbersome processes (as well as other bureaucratic complications and IP concerns) could discourage the IC from pursuing partnership with NSF—it is important to identify better ways to communicate and to build relationships so that the communities can synchronize. Regarding the issue with MOUs and other bureaucratic obstacles, Hassell suggested intermediate actions—for example, holding HHS program reviews that include government employees from agencies other than HHS. This activity protects IP and confidentiality but promotes exposure. He stressed that it is crucial to find the right people to nurture such efforts. Hassell also tries to involve HHS staff in external program reviews to observe how other agencies function and to gather ideas. Another option is to detail people temporarily to other agencies, which serves to break down cultural barriers and increase awareness. He advised that this activity would have to be done carefully within authorities. He cautioned against letting policy barriers prevent progress—policies can be changed. He added that venues such as the National Academies are valuable in bridging opportunities; when people participate in meetings sponsored by other agencies, they bring ideas back to their own organizations.

Meyerson pointed out that with the rate and pace of technology today, it is not feasible to operate as in the past and be successful. Creating a cross-agency organization that is small and highly empowered could be a viable solution. At IBM, he put together a cross-sectional team of 20 people with excellent networks who had the delegated authority to make decisions within 48 hours (instead of 6 months) about foundational changes in the direction of a program. He added that the dynamic capability to "re-aim" a program's focus is invaluable—the limitations of lengthy requirements lead to the death of innovation. He wondered if a cross-sectional approach has been attempted yet at the federal level. Hassell explained that a few similar interagency groups exist, such as the Public Health Emergency Medical Countermeasures Enterprise. However, owing to government constraints, this group still has to make its recommendation at the senior bureaucratic level.

PLANNING COMMITTEE MEMBERS Frederick R. Chang (*Chair*), Southern Methodist University; **Michael A. Marletta** (*Vice Chair*), University of California, Berkeley; **Lilian Alessa**, University of Idaho; **Tomas Diaz de la Rubia**, University of Oklahoma; **Vishva M. Dixit**, Genentech; **Donald D. Duncan**, Johns Hopkins University Applied Physics Laboratory; **Gerald L. Epstein**, National Defense University; **Kathleen Fisher**, Tufts University; **James R. Gosler**, Johns Hopkins University Applied Physics Laboratory; **Laura M. Haas**, University of Massachusetts Amherst; **Robert F. Hale**, Center for a New American Security; **Daniel E. Hastings**, Massachusetts Institute of Technology; **Frances S. Ligler**, Texas A&M University; **Willie E. May**, Morgan State University; **Bernard S. Meyerson**, IBM Corporation; **Lisa J. Porter**, LogiQ; **Peter Schiffer**, Yale University; **Anthony J. Vinci**, Center for a New American Security; **Michael S. Witherell**, Lawrence Berkeley, National Laboratory.

STAFF Dionna Ali, Associate Program Officer; **Anita Eisenstadt**, Program Officer; **Shenae Bradley**, Administrative Assistant; **Anthony Fainberg**, Senior Program Officer; **Michael Niles**, Senior Program Officer; **Nia Johnson**, Program Officer; **Alan H. Shaw**, Director.

DISCLAIMER This Proceedings of Workshop—in Brief was prepared by **Linda Casola** as a factual summary of what occurred at the workshop. The statements made are those of the rapporteur or individual workshop participants and do not necessarily represent the views of all workshop participants; the planning committee; or the National Academies of Sciences, Engineering, and Medicine.

REVIEWERS To ensure that it meets institutional standards for quality and objectivity, this Proceedings of a Workshop—in Brief was reviewed by **Dewey Murdick**, Georgetown University, **Eliahu Niewood**, The MITRE Corporation, and **Peter Schiffer**, Yale University.

SPONSORS This workshop was supported by the Office of the Director of National Intelligence.

SUGGESTED CITATION National Academies of Sciences, Engineering, and Medicine. 2022. *Leveraging the Future Research and Development Ecosystem for the Intelligence Community by the U.S. Research Community: Proceedings of a Workshop—in Brief*. Washington, DC: The National Academies Press. https://doi.org/10.17226/26605.

Division on Engineering and Physical Sciences

Copyright 2022 by the National Academy of Sciences. All rights reserved.

NATIONAL ACADEMIES Sciences Engineering Medicine

The National Academies provide independent, trustworthy advice that advances solutions to society's most complex challenges.
www.nationalacademies.org

B

Leveraging the Future Research and Development Ecosystem for the Intelligence Community—Understanding the International Aspect of the Landscape: Proceedings of a Workshop—in Brief[1]

This Proceedings of a Workshop—in Brief was published as a standalone document, so its inclusion here is for the convenience of the reader.

On June 9, 2021, the Intelligence Community Studies Board (ICSB) of the National Academies of Sciences, Engineering, and Medicine convened a virtual workshop on behalf of the Office of the Director of National Intelligence. This unclassified workshop was the second of two data-gathering workshops. Panel 1 on North Atlantic Treaty Organization (NATO)/Department of Defense (DoD) Perspectives featured panelists Bryan Wells, NATO; Captain James Borghardt, USN, ONR Global; Jason Matheny, Office of Science and Technology Policy (OSTP); and Steward Remaly, DoD/ASD/SOLIC/Irregular Warfare Technical Support Directorate. Panel 2 on U.S. Government Perspectives featured Kevin Wolf, Akin Gump LLP; Dana Johnson, OUSD(R&E); and Dewey Murdick, Georgetown University. Panel 3 on Emerging Technologies and National Security featured Richard Johnson, Global Helix, and Melissa Flagg, Georgetown University. This workshop will help to inform a consensus study report investigating how the Intelligence Community (IC) could leverage the evolving research and development (R&D) ecosystem to meet its future needs and garner the most value from its investments in science and technology (S&T). This, the second of two workshops, focused on the IC's connections with foreign R&D and approaches for protecting classified IC information while remaining abreast of cutting-edge advances (often made by foreign researchers) in critical S&T areas. The consensus study committee vice chair Michael Marletta, C.H. and Annie Li Chair in the Molecular Biology of Diseases, University of California, Berkeley, invited panelists and participants to discuss the international perspectives of how the IC could develop strategies to take advantage of this R&D landscape.

NATO AND DOD PERSPECTIVES

Bryan Wells, NATO's chief scientist, described its emphasis on emerging and disruptive technologies (EDTs), such as autonomy, artificial intelligence (AI), data, space, hypersonics, biotechnology and human enhancement, quantum technologies, and novel materials. In March 2020, NATO's Science and Technology Organization (STO) published *Science & Technology Trends 2020–2040: Exploring the S&T Edge*.[1] He underscored the value of this public release document as a way to increase dialogue among a broad range of non-NATO organizations. It provides leadership with an overview of EDTs; and it explores why EDTs are important, how they could be developed in the future, and implications for the alliance. It also highlights future opportunities for NATO, and, to the extent possible in a public

---

[1] NATO STO, 2020, *Science & Technology Trends 2020–2040: Exploring the S&T Edge*, http://www.nato.int/nato_static_fl2014/assets/pdf/2020/4/pdf/190422-ST_Tech_Trends_Report_2020-2040.pdf. Other relevant NATO publications include *2020 Highlights of the Science and Technology Organization: Empowering the Alliance's Technological Edge*; *NATO STO-CMRE: 2020 Annual Report*; and *2021 Collaborative Programme of Work: NATO's Science and Technology Organization*.

---

[1] Reprinted from National Academies of Sciences, Engineering, and Medicine. 2022. *Leveraging the Future Research and Development Ecosystem for the Intelligence Community—Understanding the International Aspect of the Landscape: Proceedings of a Workshop—in Brief*. Washington, DC: The National Academies Press. https://doi.org/10.17226/26604.

document, identifies potential threats. NATO defines the best new technologies as intelligent, digital, interconnected, and distributed, and it provides professional scientific judgments on the implications of EDTs. Wells remarked that it is more likely for a combination of technologies to create a disruptive military effect than an individual technology. The most potentially impactful combinations are AI, data, and autonomy; AI, data, and biotechnology; AI, data, and materials; data and quantum; space and quantum; and space, hypersonics, and materials. Technologies are also expected to come to military maturity on different time scales—for example, data, AI, autonomy, space, and hypersonics are expected to be disruptive within 5–10 years, while quantum, biotechnology, and novel materials are expected to be disruptive in 10–20 years. He explained that EDTs are inexpensive and easy to use, dominated by the civil sector, and spread over a large range of technical areas. This creates challenges for NATO in that the civil sector now dominates technology development, and Western societies have to find ways to surpass their adversaries. Thus, he advocated for NATO and other defense organizations to reach out to the civil sector to better understand new technologies. He expressed optimism about maintaining a technological edge, noting that 18 of the top 20 universities in the world are located in ally nations, and he highlighted the value of NATO's Technology Watch in prioritizing technology development.

Wells briefly discussed NATO 2030,[2] an initiative launched in June 2020 with an independent group of experts, emerging leaders, and representatives from civil society and the private sector. When the initiative launched, the NATO Secretary General defined its intention, within the decadal time frame, to develop further collaborations with like-minded countries (e.g., Australia, Japan, New Zealand, and South Korea) and to ensure that work on new technologies reflects the norms and standards of allied nations. Wells noted that all of NATO's innovation work is connected via the Innovation Board, which oversees EDT strategy implementation—with a focus on technology, policy, legal, and ethical issues—and is chaired by the Deputy Secretary General. Members include the Chairman of the Military Committee; the Supreme Allied Commander, Transformation; the Supreme Allied Commander, Europe; the Director General of the International Military Staff; the Assistant Secretary General on Emerging Security Challenges; and the Chief Scientist. An independent advisory group of academics, subject-matter experts, and chief executives is also included on the Innovation Board. This board looks at broader innovation priorities for NATO. It looks at innovation from many points of view and to develop strategies of implementation for emerging and disruptive technologies. The board deals with more than just technical issues; its remit also encompasses policies and legal and ethical issues resulting from this new set of technologies.

Serving as session moderator, Alan Shaw, director of ICSB, National Academies, asked how NATO characterizes success in its engagement with other countries. Wells replied that four non-ally nations (Sweden, Finland, Australia, and Japan) are engaged closely with CPoW, which maintains a low level of classification. This makes it easier for individual nations to invite non-NATO partners into projects without security barriers and to create an environment of mutual benefit. Robert Hale, adjunct senior fellow, Center for a New American Security, inquired about Technology Watch as it pertains to potential adversaries, and Wells commented that Technology Watch Cards describe the extent to which specific technologies are expected to evolve in the coming years. Given the low level of classification, these focus more on the opportunities for rather than on the threats to NATO allies. However, classified work is being conducted on counter-EDTs and other threats. Bernard Meyerson, chief innovation officer emeritus, IBM Corporation, wondered if NATO-funded schools still exist, and Wells responded that NATO offers a range of educational opportunities and facilities. For example, the NATO Defense College in Italy includes science, political, and military programs. NATO also offers specialist courses at the von Karman Institute for Fluid Dynamics in Belgium, and other members of the network conduct research seminars at universities. In response to a question from Marletta about why biotechnology is not yet mature within the military, Wells said that the biotechnology that has been developed in the laboratory or in the civilian sphere needs to be "ruggedized" to be effective on the battlefield. He believes that the transition to a proven military capability will happen, but it will take time.

Capt. James Borghardt, Commanding Officer, Office of Naval Research (ONR) Global, explained that ONR Global focuses on

---

[2] See the NATO 2030 website at https://www.nato.int/nato2030/.

fundamental research and transitioning it to ONR, the Naval Research Laboratory, and the broader Naval R&D enterprise. He noted that all of the research that ONR Global funds is open and published, although some eventually shifts to classified projects. Sharing several statistics about the state of S&T, he said that according to the 2017 *Global R&D Magazine Funding Forecast*, 80 percent of the world's researchers are outside of the United States. The Congressional Research Service's *Global R&D Expenditures 2020 Fact Sheet* reported that the U.S. share of global R&D funding decreased from 69 percent in 1960 to 28 percent in 2018. According to the *Times Higher Education World University Ranking 2021*, 25 of the top 50 universities and 63 of the top 100 universities are outside of the United States. And of the $2.1 trillion in global research funding, DoD basic research funding accounts for 0.1 percent. Thus, he asserted that the United States (and DoD in particular) is no longer driving global basic research.

One of the primary efforts within ONR Global is to better understand state-of-the-art science in both neutral and adversarial countries. Instead of investing in specific capabilities, Capt. Borghardt continued, ONR Global invests in trusted partnerships—the 0.3 percent of the DoD basic research funding that ONR Global invests is used to connect leading international researchers. The ability to create a collaborative network of research networks to accelerate discovery overrides the funding imbalance. He noted that these trusted partnerships can be difficult to establish, yet they are successful owing to ONR Global's "enduring presence." ONR Global increases trust among its partners by spending time in international laboratories, with students, and at conferences of global scientists. ONR Global's 50 scientists and engineers work in 2 offices and 25 deployment sites to transition the network of connections into a network of trusted partnerships. These relationships continue when scientists return from their deployments to the Naval R&D laboratories, federally funded research and development centers (FFRDCs), and university affiliated research centers (UARCs). In closing, he referred workshop participants to a publicly available annual prospectus[3] that details how ONR Global has allocated its time, resources, funding, and staff during the past year.

Shaw asked how the IC could gain trust with scientists around the world. Capt. Borghardt reiterated the value of establishing an enduring presence. For example, in an effort to preserve trust, ONR Global decided at the start of the pandemic that as long as individual scientists were comfortable, they would not be removed from their locations. He advised the IC that building a network of trusted partners takes time but added that the IC has ample opportunity to develop such a network. Tomas Diaz de la Rubia, vice president for research and partnerships, University of Oklahoma, questioned whether there is a forum by which ONR Global shares its findings about global S&T with the IC. Capt. Borghardt confirmed that, as one of the five active-duty military officers in ONR Global, part of his role is to ensure that topics of interest to the IC are shared via the proper channels. Peter Schiffer, vice chancellor for research and professor of physics, University of Illinois at Urbana-Champaign, wondered how ONR Global tracks the full breadth of research and maps it onto what might be relevant to ONR, the military, and the IC. Capt. Borghardt acknowledged that this is a difficult endeavor. He has been advocating for a tool to increase information sharing across networks as well as an AI-enabled knowledge management tool to improve collection and dissemination. With approximately eight scientists per region who have domain knowledge and a specific area of interest knowledge, it is possible to better collate information. However, the fact that this knowledge is managed manually is not optimal. Schiffer asked about the likelihood of deploying an AI-enabled tool that would be effective throughout the diverse global research enterprise. Capt. Borghardt responded that the U.S. Army, Navy, and Air Force support this type of tool because it would enable synergy without needing to be in the same country. He expressed cautious optimism about deploying such a tool in the future: people understand the need and the value, but it takes a long time to make changes. He emphasized that the technology is readily available in the civil sector; it is a matter of securing resources and transitioning the tool for defense use.

Jason Matheny, Deputy Director for National Security, OSTP, commented that the IC is not currently well organized (in terms of resources, tools, and expertise) to evaluate developments and trends in globally emerging commercial and dual-use technologies. This is in part because using tools for open-source

---

[3] ONR Global, 2020, *ONR Global International Science Prospectus for FY20*, https://www.onr.navy.mil/en/Science-Technology/ONR-Global/About-ONR-Global.

intelligence can be unwieldy on the IC's classified systems. At Georgetown University's Center for Security and Emerging Technology (CSET), it is possible to do such analysis at scale in an open setting with access to the university's modern data science tools and expertise. He advocated for the scaling up of CSET to cover more technical areas, perhaps by leveraging the network of UARCs, where students and faculty are already engaged in relevant research and competitive analysis of the S&T landscape. He championed drawing not only on the competitive intelligence of academia but also on the competitive intelligence of the commercial sector. The challenge in both cases is that it can be difficult for the IC to reach out directly to those universities and companies without a contract in place (i.e., multiple layers of approval are needed). Matheny suggested creating a clearinghouse that is separate from but accessible to the federal government to facilitate interactions with these academic and commercial communities, which could lead to more timely, cost-effective, and accurate intelligence on global S&T.

Matheny expressed his hope that the more than $100 billion expected from the Creating Helpful Incentives to Produce Semiconductors for America Act and the Endless Frontier Act would create more opportunities for S&T funding. However, he said that new mechanisms for spending those funds would be useful. He remarked that although both grants and contracts have important roles within government S&T funding, grants, for instance, do not clearly specify desired research results and make it difficult to measure progress. He advocated for conducting experiments in *how* S&T is funded as a way to develop more cost-effective, higher performing mechanisms (e.g., prize challenges, regulatory review vouchers, advanced market commitments, advanced purchase commitments). He noted that it is also important to explore how the government makes decisions about who receives these awards. Most rely on a group judgment of experts, but such group judgments are prone to error. Using the research on forecasting could help better anticipate which proposals are likely to yield the most valuable scientific research. Another approach, he continued, would be to use randomized awards. He suggested requesting proposals on innovative decision making for funding science and creating a controlled experiment, in which all proposals are funded, although via different mechanisms. Five years later, a determination could be made about which research projects were most successful.

Matheny underscored the value of conducting more science on the funding of science.

Schiffer wondered if the funding agencies are willing to consider alternative funding mechanisms. Matheny credited the Defense Advanced Research Projects Agency (DARPA) for its creative approaches both in soliciting ideas from the research community and to funding them. He added that the National Science Foundation is interested in experimenting with new approaches, and he is hopeful that people will have top-cover from OSTP and the Office of Management and Budget to experiment with the tools of science funding. Shaw asked Wells if NATO has discussed alternative approaches to funding innovation. Wells noted that while non-traditional funding mechanisms have been considered, NATO has not identified any that would be appropriate. He said that funding mechanisms that work at a national level may not work at an international level, owing to different intellectual property rules and competition rules and thresholds throughout the international community. And when there are 30 allies and decisions are taken by consensus, the desire to take on risk decreases. Marletta observed a general reluctance to change the scientific peer-review process, despite the fact that the approach is not always effective. He supported Matheny's idea to have a controlled experiment on improved proposal assessment and expressed his hope that the academic community would embrace it. Matheny added that opportunities are missed when almost all of the science organizations in the federal government make funding decisions using nearly identical processes.

Meyerson observed that private-sector companies use a variety of strategies to remain competitive, and he wondered if studies have been conducted on the longevity of the 20 most successful companies. Following that line of thought, Matheny questioned whether there have been studies on technologies that accelerate the overall rate of innovation within a society as well as whether there is enough investment in such technologies, as the government tends to fund individual research projects instead of clusters of research projects that could develop a tool to elevate an industry. Matheny highlighted the importance of better understanding pre-competitive industry consortia that could enable the acceleration of such tools. Meyerson mentioned the challenge of convincing the IC that leveling the playing field (via open-source work) would be better than being left behind by its competitors.

Steward Remaly, Program Manager for Surveillance, Collection, and Operations Support, Irregular Warfare Technical Support Directorate (IWTSD), described IWTSD as an organization of "experienced professionals" under the umbrella of the Assistant Secretary of Defense, Special Operations/Low Intensity Conflict. Its mission is to identify and develop capabilities for DoD to conduct irregular warfare against all adversaries, including Great Power competitors and non-state actors, and to deliver those capabilities to DoD components, the interagency, and foreign partners through rapid R&D, advanced studies, and technical innovation. He emphasized that IWTSD's work spreads to IC partners. Its objectives for the upcoming year are to support the National Defense Strategy and the Annex for Irregular Warfare; provide forums to solicit and collaborate on R&D requirements; rapidly advance technology development, deliver prototypes for operational tests and evaluations, and assist in product transition; promulgate technology and information exchange; and influence policy and identify enablers. Remaly asserted that new processes and increased speed of innovation make it possible to remain relevant in the R&D business. IWTSD increasingly relies on AI, big data analytics, and machine learning as it works to strengthen and increase partnerships to solve difficult problems. IWTSD aims to enhance lethal capabilities and to improve survivability, for example, within 1–2 years of receiving a requirement instead of within 5–30 years. IWTSD is leading more than 269 research, development, test, and evaluation projects, including 128 International Task Plans—maintaining memoranda of understanding (MOUs) with Australia, Canada, Israel, Singapore, and the United Kingdom. He explained that these strategic partnerships drive innovation in that they allow 50 percent cost sharing and 50 percent capability sharing. While some of the projects within the International Task Plans enable information sharing, others turn into prototypes that are fielded and become programs of record.

IWTSD has a user-focused business process, Remaly continued. No action is ever taken without a validated requirement, which comes from end users, sponsors, and subject-matter experts in an effort to drive innovation. IWTSD then turns a requirement into a broad agency announcement (BAA), which is distributed to industry partners. After a series of white papers and proposal reviews, contract negotiations begin several months later. For example, IWTSD distributed 45 requirements in February 2021 to industry via the fiscal year (FY) 2022 BAA and hopes to release funding in October 2021 (IWTSD operates on a $70–100 million annual contract). He detailed common IWTSD challenges for FY 2022: vehicle telematics data retrieval; multi-role, field-configurable, offensive small unmanned aerial system; unmanned aerial vehicle payload for rapid chemical plume detection, identification, and mapping; casualty tracking and monitoring system; and advanced cyber physical test bed. Specific opportunities for collaboration include Challenge Programs (e.g., U.S./Australia Counter Improvised Threat Grand Challenge; U.S./Israel Mobile Standoff Autonomous Indoor Capabilities Challenge; U.S./U.K. Chemical Munitions Destruction Industry Challenge) and Planned Information-Sharing Workshops (e.g., U.S./U.K. Exploiting Emerging Commercial Space, National Aeronautics and Space Administration Robotics Workshop, U.S./Israel Less Than Lethal Capabilities, U.S./U.K. Maritime Security Data Exchange).

Gerald Epstein, distinguished research fellow, Center for the Study of Weapons of Mass Destruction, National Defense University, posed a question about the role of U.S. export controls. Remaly replied that while MOUs allow IWTSD to share capabilities across defense industries, if future development or production will require International Traffic in Arms Regulations (ITAR) approval, IWTSD has to work through all of the agreements that the vendor has to work through. Dewey Murdick, interim director, CSET, Georgetown University, asked how efforts are prioritized within IWTSD. Remaly explained that during the requirements generation, collaborative discussions with end users, other R&D organizations, and IWTSD's team of experienced professionals occur, which include consideration for what others are doing, whether an idea makes sense, if something similar did not work in the past, and potential long-term ramifications. However, Remaly stated that prioritization is ultimately determined by user need; if the need is critical, it will be addressed.

PERSPECTIVES FROM INSIDE THE U.S. GOVERNMENT

Kevin Wolf, partner, Akin Gump LLP, explained that after World War II, military- or intelligence-related material required a license, no matter the level of sensitivity or relationship with the recipient. He said that ITAR hindered national security, owing to the information-sharing limitations it placed on allies. In August 2009, President Obama directed the agencies involved in the U.S. export control system to conduct a review of export controls to identify strategies to enhance U.S. national security, and, in 2010, former Secretary of Defense Robert Gates's policy objectives were as follows: reform the export control system to increase interoperability with NATO and other close allies, reduce the current incentives for companies in non-embargoed countries to design out or avoid U.S.-origin content, and allow the administration to focus its resources on the transactions of greater concern. Wolf added that spending so much time and money on reviewing and clearing trade with NATO allies reduced the time available to focus on adversaries. U.S. agencies then began reviewing the U.S. Munitions List (USML) to determine what items no longer warranted control. To implement former Secretary Gates's objectives, Wolf continued, "the administration needed to identify the specific sensitive and other items on a more positive USML that warrant individual license reviews even for ultimate end use by NATO and other regime allies; and amend the Export Administration Regulations and the Commerce Control List to control all formerly USML items that would no longer be on the revised USML so that they still could be adequately controlled but in a more flexible way regarding such allies." The embargo on China remained, as did significant limitations for Russia and complete embargoes on Iran, North Korea, and Cuba. The Strategic Trade Authorization, a liberal license exception, allowed for the sharing of commodities, technology, and software among nationals and governments of the allied countries. Wolf noted that many of the descriptions in the regulations pertaining to things of concern to the IC were deliberately vague; his team worked with the agencies to describe these in more detail, but barriers remain.

Serving as session moderator, Epstein pointed out that many of the technologies of interest to the IC will not be developed by IC partners; collaboration is one of the best ways to understand global developments, and two-way interaction is necessary to gain the most benefit from partnerships. He observed that export controls intersect with the issues of the IC, and he asked Wolf to address strategies to deal with high-tech companies overseas. Wolf said that, motivated by changes in Chinese technology acquisition efforts, civil–military fusion policies, and human rights abuses, an important issue now concerns the AI, robotics, nanotechnology, and quantum computing technologies that are not on the export control list but warrant control. He added that the international export control system is not built to address the kinds of threats and novel technology acquisition efforts of the commercial space. In 2018, Congress passed more expansive authority for the regulating agencies. The Department of Commerce followed up with an interagency effort to identify emerging technologies that are not controlled but are essential to national security, and the Biden administration is defining a new concept of national security. The challenge is that U.S. allies' systems embrace the previous export control mindset of only regulating weapons systems–related items. Wolf commented that a significant amount of rethinking is required on what warrants control to prevent technology of concern from going to China or Russia, without limiting it among the allies.

Dana Johnson, Director, International Outreach and Policy (IO&P), Office of the Under Secretary of Defense (OUSD)/Research and Engineering (R&E), explained that IO&P emerged in 2018 as the focal point for all international S&T[4] engagement activities within OUSD(R&E). IO&P advises USD(R&E) and supports OUSD(R&E) on defense strategy, policy, and engagement for international matters. IO&P works closely with DARPA, the Defense Innovation Unit, the Missile Defense Agency, the Space Development Agency, the Defense Technology Information Center, and the Test Resource Management Center; provides subject-matter expertise; and assists OUSD(R&E) in pursuing international S&T cooperative activities aligned with the DoD Modernization Priorities.[5]

In December 2020, DoD *International Science and Technology Engagement Strategy: A Unified Approach to Strengthen Alliances and Attract New Partners (ISTES)*[6] was released, providing a rationale for engaging with other countries. Johnson remarked that the mission of ISTES is to leverage foreign defense S&T capabilities, develop relationships with other countries to access these capabilities, maximize coalition interoperability, and achieve U.S. national security objectives. The document provides strategic guidance and coordinates individual DoD Component

---

[4] Johnson defined S&T as basic and applied research through experimentation and prototyping.

[5] The DoD Modernization Priorities focus on AI/machine learning; fully networked command, control, and communications; quantum science; autonomy; space; biotechnology; microelectronics; directed energy; cyber; hypersonics; and 5G.

[6] DoD, 2020, *Department of Defense International Science and Technology Engagement Strategy: A Unified Approach to Strengthen Alliances and Attract New Partners*, https://www.cto.mil/wp-content/uploads/2020/12/Signed-International-ST-Engagement-Strategy.pdf.

engagement activities toward common objectives; directs international outreach efforts toward opportunities with the highest return on investment; seeks new opportunities with friendly nations that are pursuing niche S&T capabilities of interest to the United States; and provides a structured approach for creating enhanced awareness, coordination, and strategic planning of defense international S&T engagements. ISTES does not direct which nations to partner with or what S&T to pursue because S&T priorities may change over time, and alliances and regional interests may affect those choices. The vision of ISTES, she continued, is to have consistently deliberate engagement with allies and partners; share awareness of global sources of technology, which enables identification of gaps or duplication of efforts as well as any priorities subject to technology protection or controls; create well-established international relationships and effective mechanisms for cooperation (e.g., using existing agreements to share information); have visible senior leaders to engage with foreign counterparts and steer cooperation; create thriving international networks of researchers with collaborations in priority S&T areas producing high-volume outcomes; and maintain continuous improvement to business processes. She explained that ISTES principles

- align with U.S. interests, including national security objectives, specific S&T needs, and international policy;

- balance strengthening existing alliances with forging new partnerships to secure U.S. access to world-class S&T and achieve desired posture;

- prioritize S&T investment and resources according to U.S. S&T needs, foreign S&T strengths and opportunities, and U.S. policy;

- protect security of critical U.S. technologies;

- are justified with benefits that exceed those that could be achieved independently;

- are equitable for all parties;

- strive for measurable outcomes that accelerate the pace of U.S. R&D and ultimately benefit defense missions;

- are underpinned by suitable agreements and arrangements offering flexibility; and

- enable timely and effective interactions that accommodate foreign governments' requirements.

In response to a question from Epstein about specific partnerships with other countries, Johnson highlighted The Technical Cooperation Program (TTCP) with the Five Eyes partners, which has several ongoing activities and has been active for 60 years. Another example is a partnership with STO, for which the United States has three principal members, one of whom is a voting member from the Services. IO&P has also used the STO databases to track U.S. participation in global technology collaborations and determine where to increase U.S. engagement. Epstein asked how gaps are identified and R&D investment targeted toward those gaps. Johnson replied that the first step is talking to counterparts in the embassies and overseas. IO&P relies on open-source and government documents, and an OUSD(R&E) strategic analysis cell links to the IC, helping to determine whether a potential partner is the right fit based on mutual priorities. She expressed her hope for increased IC engagement in the future. Schiffer commented that the IC is less centralized in its organizational structure than DoD. He questioned how to navigate bureaucratic challenges and make connections with the IC. Johnson said that it could be challenging but emphasized the value of leveraging that relationship. She proposed the creation of an international community of interest and welcomed suggestions about strategies to gain better insight for decision making.

Murdick explained that CSET helps policy makers understand the security implications of emerging technologies by connecting them to high-quality analysis. Two efforts launched by the Intelligence Advanced Research Projects Activity (IARPA)—Foresight and Understanding from Scientific Exposition (FUSE) in 2011 and Forecasting S&T (ForeST) in 2014—help analysts target their attention within an expansive space and detect and forecast new technical capabilities. He emphasized that the IC plays only a minor role in R&D; however, it has an important role in monitoring threats and opportunities that could be relevant to national security.

Murdick commented that FUSE was created to achieve validated, early detection of technical emergence. It aimed to reduce technical surprise by developing reliable forecasts and indicators from English and Chinese scientific and patent literature. He stressed that complex indicators were helpful for context but were not useful in terms of predicting what would disrupt technology; the simplest indicators were the most predictive. He pointed out that ForeST picked up where FUSE left off, taking scientific- and patent-type indicators and asking the crowd for other future S&T milestones. FUSE and ForeST are now being used by CSET to explore the following questions: What fundamental innovations are coming fastest, will have the most impact (e.g., advancements, capability disruption, or performance changes in mission critical systems), and need the most attention? How can those be prioritized to respond in time? Which basic science AI-related research areas will become emerging (security-related) technologies? Who should we listen to and from which interest community? In what areas do the United States and its allies have an advantage over China? He emphasized that these are very difficult questions to answer, but this forecasting technology makes it possible to determine which technologies are likely in the next few years to grow quickly. Understanding where the attention should be focused and connecting the data from these methods to top-down strategic context supports decision making—without context, the data are not useful.

Murdick highlighted that FUSE, in its new format at CSET, takes approximately 240 million research articles and groups them into problem- and citation-based clusters. Approximately 126,000 research problem–related clusters represent defined technical areas of research and have at least 50 papers each. Next, extreme growth research cluster forecasts are done at scale (selecting the top 3–4 percent most quickly growing clusters, estimated to grow at least 8 percent per year for the next 3 years) to begin to understand and prioritize which are growing most quickly and are connected to military interests. Multiple filters and linkages can then be used (e.g., for military interest, sectoral/corporate interest, news, social media). He described this approach as evidence-driven down-selection, and he emphasized the value of this tool for the IC—which would never have enough resources or expertise to track and understand the research—to contextualize and determine what could hinder competitive advantage. An important lesson learned, he continued, is that forecasting is useful for discriminating and helping eliminate unhelpful models; simple models perform better forecasting than complex models. Describing ForeST (now referred to as Foretell), he noted that IARPA's Aggregative Contingent Estimation program established the accuracy of crowd forecasting on *well-defined questions*. Challenges arose for "big questions" relevant to policy makers. Foretell's solution was to combine expert and crowd judgment: use expert judgment to break policy issues down into forecastable questions, elicit crowd forecasts on the questions about which experts disagree or are uncertain, and aggregate the results across questions to provide actionable insight. In closing, Murdick shared several insights: human judgments about technical emergence do not make for effective ground truth, owing to poor temporal resolution and linearization of memories; subject-matter expert "favorites" persist; and small-group subject-matter expert forecasting often has a low level of accuracy. He explained that technology transfer to the IC is particularly challenging because the IC is not well organized for using open data sources, and reorganizations change S&T analysis resourcing.

Schiffer asked if Murdick's team has conducted retrospective studies on technologies of relevance to the IC to understand whether the emergence of these technologies could have been predicted if forecasting technology had been applied. Murdick said that approximately 30 case studies were performed on relevant emerging technologies over the past 20–30 years. While it was possible to validate methods, it was difficult to find methods that were discriminative. He added that most retrospective studies are not helpful because they provide a false sense of achievement; the reality of the future is what matters most. Murdick noted that there is more work to be done in making successful predictions: approximately 87 percent of what was forecasted in 2014 materialized in 2017, and of the things that achieved that threshold, approximately 48 percent had been successfully predicted. Epstein observed that the data points Murdick described were only from publications and patents, and inquired about gathering and analyzing unpublished developments from companies. Murdick commented that the bulk of basic research emerges in publications and patents, but there is a dearth of information in terms of process methods, because one has to wait for a product to emerge.

Marletta posed a question about addressing the IC's challenges with technology recognition and transfer. Murdick noted the importance of accepting that the IC has limited competency to tailor and implement highly technical capabilities. Clear communication of expectations would be beneficial; and demonstrations, use cases, and defined problem sets would also be helpful in attracting "champions." He added that the IC's organizational constructs could be changed to place greater value on innovations created outside of the IC. Marletta also asked about ways to confront the challenge of the structural reorganizations that result in a loss of champions. Murdick remarked that, in many cases, S&T staff continually rotate between being put together and being dispersed among mission groups. The best approach is to implement a solution that is robust to this changeability. He emphasized that reorganization is unavoidable in any institution, but because of the IC's constant reorganizations, he expressed concern about its ability to perform R&D threat assessments at the necessary scale.

EMERGING TECHNOLOGIES AND NATIONAL SECURITY: EXAMPLES OF THE CONNECTIONS

Richard Johnson, chief executive officer, Global Helix, emphasized that 21st-century innovation will rely increasingly on biology and suggested that the IC increase its awareness of and engagement with R&D developments in synthetic biology and engineering biology (SB/EB) in particular. He explained that SB/EB has the potential to harness the intrinsic capabilities of biological systems to build life-like systems to explore the mechanisms that govern biology; manufacture products that are safer, more sustainable, and environmentally friendly; improve human health; develop computing and information storage devices; and be used as active materials that sense and respond to their environment. More than 40 countries have SB/EB strategies and R&D programs, the most comprehensive and advanced of which are in China, the United States, and the United Kingdom.

Johnson emphasized that the IC could monitor investments in cutting-edge SB research. For example, SynBioBeta releases quarterly market reports, projecting $36 billion in new investment for SB companies in the United States this year. And *The Bio Revolution: Innovations Transforming Economies, Societies, and Our Lives*[7] anticipated that 60 percent of the world's physical inputs could be made using biological means as well as a projected $2–4 trillion of annual direct economic potential globally for SB in 2030–2040. He added that the IC could track university resource allocation trends (e.g., new centers, faculty hiring, emerging institutes) and international funding trends (e.g., Shenzhen Institute of Synthetic Biology). The IC could also build, deploy, and iterate a new robust analytical toolkit for SB/EB R&D, Johnson continued. For example, the Engineering Biology Research Consortium (EBRC) offers a critical assessment of the potential of EB and related roadmaps,[8] with contributions from more than 90 scientists and engineers from a range of disciplines, representing more than 30 universities and 12 companies. Developing next-generation analytical mapping and modeling tools is key, he asserted, as is focusing more on anticipatory intelligence and the changing landscape of SB (via horizon scanning and improved identification of novel risk pathways). He suggested that the IC reframe its mindset for next-generation SB/EB, broadening the scope of U.S. national security interests.[9] In the midst of a tool revolution, SB is increasingly focused on the automation of biology and workflows, use of AI and new data, and digitalization of biology. He maintained that if the IC does not focus on these tools and applications, it will miss many opportunities. Biodesign, the notion of designing R&D to achieve particular functionalities or outcomes, is also important for the IC. He referenced the Global Biofoundries Alliance as an example of next-generation critical infrastructure that the IC could consider. It is crucial, he continued, for the IC to leverage biology as a technology platform for multiple transformative applications. This drives the next production revolution, which will have major implications for U.S. economic growth and for the competitiveness of its manufacturing base. As the IC expands the focus of its landscape (e.g., bioinformatics, neurotechnologies, gene editing, engineering microbiomes), he indicated that it is also important to consider risk mitigation and to focus more strategically on dual-use technologies. Johnson explained that there are estimates that 40 percent of the next generation of storage computing and semiconductor technologies will be bio-based, and it is crucial that the IC keeps track of those next-generation technologies.

Johnson underscored the value of a more coordinated, whole of government strategy for critical emerging technologies in R&D. For example, interagency workshops on SB could be integrated

---

[7] McKinsey Global Institute, 2020, *The Bio Revolution: Innovations Transforming Economies, Societies, and Our Lives*, https://www.mckinsey.com/industries/pharmaceuticals-and-medical-products/our-insights/the-bio-revolution-innovations-transforming-economies-societies-and-our-lives.

[8] See the EBRC Roadmaps website at https://roadmap.ebrc.org.

[9] Johnson referred participants to the following publications: National Research Council, 2009, *A New Biology for the 21st Century*, Washington, DC, The National Academies Press; National Research Council, 2014, *Convergence: Facilitating Transdisciplinary Integration of Life Sciences, Physical Sciences, Engineering, and Beyond*, Washington, DC, The National Academies Press.

with both the IC and the defense community. A proactive strategy could be implemented for international efforts, and an enterprise similar to the six-academies initiative[10] could be developed to provide insights.

Serving as session moderator, Tony Fainberg, senior program officer, National Academies, asked how EBRC is addressing security. Johnson described Malice Analysis, a program that is gauging how university students think about risk, and noted a broad range of activities in security; some are funded by the Department of Homeland Security or DoD, and others are funded by non-governmental organizations.

Melissa Flagg, senior fellow, CSET,[11] Georgetown University, explained that all of the IC and DoD institutions that focus on science were created and optimized in a fundamentally different world than exists today. Now, the United States and China split approximately 50 percent of the $2.2–2.4 trillion of annual R&D globally; $1–1.2 trillion of R&D funding comes from the rest of the world. This demonstrates that S&T and R&D are not bilateral issues, which creates a challenge for the IC. Many countries are doing high-quality work in niche areas, yet the United States makes the error of discounting them because they lack a comprehensive portfolio of R&D and R&D funding. Flagg noted that CSET's global models are invaluable in providing context to understand where technologies are unique or are emerging quickly and require quick decisions (e.g., awareness of China's activities is insufficient when 40 other countries are engaging in the same activities)—a non-contextualized focus on adversarial S&T is untenable.

Flagg pointed out that it is also difficult for U.S. institutions to adjust to the new standard of highly collaborative science, which has enabled small English-speaking nations to increase their presence and capabilities. The United States often asks its allies to make the same choices it is willing to make, but it is important to recognize that such choices could hurt those allies. She explained that in the United States, not only is 78 percent of science now funded by non-governmental entities but also 90 percent of science is performed outside of the government, and a similar trend is apparent across the world. Thus, she said that it is important for the United States to rethink structures for forming alliances, improve engagement with and understanding of the domestic R&D landscape, and reconsider the IC's overall approach. Flagg underscored that the United States is unaccustomed to building partnerships in R&D where it needs something from another country. Because of this mindset, the United States often duplicates or ignores research. This presents a challenge for DoD and the IC, she continued, especially in terms of emerging technologies. She reiterated Murdick's assertion that internal experts are not necessarily the best people to validate research; novel structures would be useful.

Flagg noted that the level of internationalization and collaboration across topics of science varies dramatically—for example, China has a low level of collaboration for material science but a high level for energy and the environment. When the United States enters into conversation with potential partners, it is important to understand not only the risk but also the risk of protective choice that the United States is asking them to make. She asserted that the IC often overlooks lost opportunity costs and attempts to eliminate all risks, which is an unrealistic perspective. Furthermore, "military-only" technologies are becoming irrelevant, as most technologies have dual/commercial use. Novel approaches to consortia are one way to address these challenges. She suggested, however, that if the United States is committed to developing fruitful partnerships, neither DoD nor the national security apparatus should lead, because the allies' international S&T strategies focus on the economy instead of on national security. She cautioned that if the United States does not understand its allies, it will not be able to understand its adversaries or the broader S&T landscape.

Flagg described recent CSET research on the top 50 global defense companies by revenue, which comprise about $500 billion of annual revenue from global militaries alone. Investment activity and mergers and acquisitions activity were tracked using Crunchbase, and relevant data sets were used to study disclosed

---

[10] See National Academy of Engineering and National Research Council, 2013, *Positioning Synthetic Biology to Meet the Challenges of the 21st Century: Summary Report of a Six Academies Symposium Series*, Washington, DC, The National Academies Press.

[11] Flagg shared the following links to expanded explanations of CSET's efforts: https://cset.georgetown.edu/publication/global-rd-and-a-new-era-of-alliances/, https://cset.georgetown.edu/publication/research-security-collaboration-and-the-changing-map-of-global-rd/, https://cset.georgetown.edu/publication/comparing-the-united-states-and-chinas-leading-roles-in-the-landscape-of-science/, https://cset.georgetown.edu/publication/the-public-ai-research-portfolio-of-chinas-security-forces/, and https://cset.georgetown.edu/publication/tracking-ai-investment/.

investments in AI companies. When looking through the lens of investment, a different S&T landscape emerges. In closing, she mentioned that she is more knowledgeable of emerging technology today at CSET than she was during her tenure as Deputy Assistant Secretary of Defense for Research, in part because the IC fails to leverage open-source data.

Schiffer asked how Flagg would ensure that the Pentagon and the IC gain a better understanding of the S&T landscape. Flagg stated that ONR and the Air Force Research Laboratory (via the Air Force Office of Scientific Research) are the only institutions that have created a structure to support ongoing access to data for decision making. However, these data are rarely used to contextualize decisions by leadership. The basic research literature cannot be used to answer every question, and a misunderstanding of diffusion problems, in particular, persists. CSET has access to 90 percent of the global literature, as well as patents, investment data, on-staff translators, and survey capabilities; this suite of lenses with which to look at technology allows CSET to answer a more distinct set of questions. She asserted that governmental organizations simply need to commit to investing in and providing ongoing support of open-source efforts. Fainberg asked how DoD and the IC could engage in collaborative R&D while maintaining situational awareness and protecting items of critical interest. Johnson highlighted opportunities for collaborations with allies, and he suggested adapting the In-Q-Tel model on a global scale. He also advocated for engaging with key players in emerging technologies who are outside of the government. Flagg commented that it is important to have traditional institutional relationships as well as awareness of the fact that the vast majority of global R&D is disconnected from government. She suggested funding a seat at the table in a consortium. Most importantly, if the United States wants to become more knowledgeable of global developments, it has to enter the conversation in a way that benefits partners and does not create an opportunity cost (e.g., via export controls and other security measures). The United States has to begin to articulate access to data sets, infrastructure at laboratories, and other opportunities that it can offer potential partners, she continued. Fainberg wondered if opportunities exist for research in unclassified areas at the international level. Flagg remarked that there are many applied opportunities (e.g., manufacturing, productization, and investment pooling) with the Five Eyes if the United States relaxes its export controls. She reiterated that the United States has to be realistic about the choices it is asking partners to make.

PLANNING COMMITTEE MEMBERS **Frederick R. Chang** (*Chair*), Southern Methodist University; **Michael A. Marletta** (*Vice Chair*), University of California, Berkeley; **Lilian Alessa**, University of Idaho; **Tomas Diaz de la Rubia**, University of Oklahoma; **Vishva M. Dixit**, Genentech; **Donald D. Duncan**, Johns Hopkins University Applied Physics Laboratory; **Gerald L. Epstein**, National Defense University; **Kathleen Fisher**, Tufts University; **James R. Gosler**, Johns Hopkins University Applied Physics Laboratory; **Laura M. Haas**, University of Massachusetts Amherst; **Robert F. Hale**, Center for a New American Security; **Daniel E. Hastings**, Massachusetts Institute of Technology; **Frances S. Ligler**, Texas A&M University; **Willie E. May** (until May 2021), Morgan State University; **Bernard S. Meyerson**, IBM Corporation; **Lisa J. Porter**, LogiQ; **Peter Schiffer**, Yale University; **Anthony J. Vinci**, Center for a New American Security; **Michael S. Witherell**, Lawrence Berkeley National Laboratory.

STAFF Dionna Ali, Associate Program Officer; **Anita Eisenstadt**, Program Officer; **Shenae Bradley**, Administrative Assistant; **Anthony Fainberg**, Senior Program Officer; **Michael Niles**, Senior Program Officer; **Nia Johnson**, Program Officer; **Alan H. Shaw**, Director.

DISCLAIMER This Proceedings of a Workshop—in Brief was prepared by **Linda Casola** as a factual summary of what occurred at the workshop. The statements made are those of the rapporteur or individual workshop participants and do not necessarily represent the views of all workshop participants; the planning committee; or the National Academies of Sciences, Engineering, and Medicine.

REVIEWERS To ensure that it meets institutional standards for quality and objectivity, this Proceedings of a Workshop—in Brief was reviewed by **Bernard Meyerson**, IBM Corporation; **Melissa Flagg**, Flagg Consulting, LLC; and **Peter Sharfman**, The MITRE Corporation.

SPONSORS This workshop was supported by the Office of the Director of National Intelligence.

SUGGESTED CITATION National Academies of Sciences, Engineering, and Medicine. 2022. *Leveraging the Future Research and Development Ecosystem for the Intelligence Community—Understanding the International Aspect of the Landscape: Proceedings of a Workshop—in Brief*. Washington, DC: The National Academies Press. https://doi.org/10.17226/26604.

Division on Engineering and Physical Sciences

Copyright 2022 by the National Academy of Sciences. All rights reserved.

NATIONAL ACADEMIES Sciences Engineering Medicine

The National Academies provide independent, trustworthy advice that advances solutions to society's most complex challenges.
www.nationalacademies.org

C

Acronyms and Abbreviations

AFOSR	Air Force Office of Scientific Research
AI	artificial intelligence
CIA	Central Intelligence Agency
CRADA	Cooperative Research and Development Agreement
CSIS	Center for Strategic and International Studies
CTIO	chief technology and innovation officer
CUI	controlled unclassified information
DARPA	Defense Advanced Research Projects Agency
DDNI	Deputy Director of National Intelligence
DHS	Department of Homeland Security
DNI	Director of National Intelligence
DoD	Department of Defense
DOE	Department of Energy
D/S&T	Director for Science and Technology (ODNI)
FAR	Federal Acquisition Regulations
FFRDC	federally funded research and development center
FVEY	Five Eyes
FY	fiscal year
IARPA	Intelligence Advanced Research Projects Activity
IC	Intelligence Community
IP	intellectual property
IPA	Intergovernmental Personnel Act
ITAR	International Traffic in Arms Regulations

JDA	joint duty assignment	
NASA	National Aeronautics and Space Administration	
NATO	North Atlantic Treaty Organization	
NGA	National Geospatial-Intelligence Agency	
NIH	National Institutes of Health	
NIM	National Intelligence Manager	
NIO	National Intelligence Officer	
NISTC	National Intelligence Science and Technology Committee	
NITRD	Networking and Information Technology Research and Development	
NNSA	National Nuclear Security Administration	
NRO	National Reconnaissance Office	
NSA	National Security Agency	
NSF	National Science Foundation	
NSS	National Security Strategy	
NSTC	National Science and Technology Council	
ODNI	Office of the Director of National Intelligence	
ONR	Office of Naval Research	
OSINT	Open Source Intelligence	
OSTP	Office of Science and Technology Policy	
R&D	research and development	
RFP	request for proposals	
S&T	science and technology	
S&TI	science and technology intelligence	
TDY	Temporary Duty	
TRL	technology readiness level	
UARC	university affiliated research center	
USGCRP	U.S. Global Change Research Program	

D

Committee Member Biographical Information

FREDERICK R. CHANG, *Chair*, is the chair of the Computer Science Department in the Lyle School of Engineering at Southern Methodist University (SMU). He is also the Bobby B. Lyle Endowed Centennial Distinguished Chair in Cyber Security and a professor in the Department of Computer Science. He is the founding director of the Darwin Deason Institute for Cyber Security and a senior fellow in the John Goodwin Tower Center for Public Policy and International Affairs in SMU's Dedman College. Additionally, Dr. Chang's career spans service in the private sector and government, including as the former director of research at the National Security Agency. Dr. Chang was elected as a member of the National Academy of Engineering (NAE) in 2016. Until recently he served as the co-chair of the Intelligence Community Studies Board of the National Academies of Sciences, Engineering, and Medicine, and he is also a member of the National Academies' Army Research Laboratory Technical Assessment Board. He has served as a member of the National Academies' Computer Science and Telecommunications Board and as a member of the Commission on Cybersecurity for the 44th Presidency. He is the lead inventor on two U.S. patents and has appeared before Congress as a cybersecurity expert witness on multiple occasions. Dr. Chang received his B.A. from the University of California, San Diego, and his M.A. and Ph.D. from the University of Oregon. He has also completed the Program for Senior Executives at the Sloan School of Management at the Massachusetts Institute of Technology (MIT). He has been awarded the National Security Agency Director's Distinguished Service Medal.

MICHAEL A. MARLETTA, *Vice Chair*, is a professor of chemistry, a professor of molecular and cell biology, and the CH and Annie Li Chair in the Molecular Biology of Diseases at the University of California (UC), Berkeley. After an A.B. in biology and chemistry from SUNY Fredonia in 1973, he received a Ph.D. in 1978 from University of California, San Francisco (UCSF), followed by a 2-year postdoctoral appointment at MIT. In 1980, Dr. Marletta joined the faculty at MIT. In 1987, he moved to the University of Michigan and in 1991 was appointed as the John G. Searle Professor of Medicinal Chemistry Chair. In 1997, he became an investigator in the Howard Hughes Medical Institute. Dr. Marletta moved to UC Berkeley in 2001, where he assumed the positions of professor of chemistry, Department of Chemistry, and professor of biochemistry and molecular biology, Department of Molecular and Cell Biology. He was appointed the Aldo DeBenedictis Distinguished Professor of Chemistry in 2002. He served as the chair of the Department of Chemistry at UC Berkeley from 2005 to 2010. In 2011, he joined the faculty of the Scripps Research Institute and was named the president-elect. He assumed the presidency in January 2012. He returned to UC Berkeley in 2015. He recently became the co-chair of the National Academies'

Intelligence Community Studies Board. He is an elected member of both the National Academy of Sciences (NAS) and the National Academy of Medicine (NAM).

LILIAN ALESSA received her Ph.D. from The University of British Columbia in 1998 and is the President's Professor and the director of the Center for Resilient Communities at the University of Idaho. She holds a research affiliate position with the Bush School of Government and Public Service at Texas A&M University and has served at the Defense Intelligence Senior Level, through an Intergovernmental Personnel Act, with the Department of Defense (DoD). She was also the deputy chief of global strategies at the Department of Homeland Security (DHS), Office of Strategy, Policy and Plans, where she contributed to the department's strategic development and implementation of an estimated $16 billion budget for combined strategies. In addition to her public service, she has been, or is, a member of the board of directors for several national initiatives such as the Arctic Research Consortium of the United States and the National Ecological Observatory Network. She served two terms on the National Science Foundation (NSF) Advisory Committee for Environmental Research and Education, which reported directly to the Director of NSF. She co-authored the NSF decadal plan titled "America's Future: Environmental Research and Education for a Thriving Century" and together with her colleagues at Arizona State University she co-founded the Community of Modelers in Social Ecological Systems, an international community and cyberinfrastructure of researchers, educators, and professionals dedicated to the transparency, reproducibility, and best practices for advanced agent-based modeling. Based on her 25 years of experience, she continues to act as an academic advisor to senior leaders across a range of federal, state, local, tribal, and territorial agencies on complex topics that bridge security, intelligence, and defense with the resilience of systems, people, and communities. She has led several multi-million-dollar integrative science and data fusion programs working across the intelligence, law enforcement, and defense communities; academia; and public stakeholders particularly in the areas of all-source intelligence, natural resource security, social stability, and collaborative resilience. Trained as both a social and physical scientist she has worked to ensure that appropriate technologies are evaluated and operationalized within the constructs of the sociocultural systems they serve. In addition to managing a $128 million academic portfolio, publishing more than 100 peer reviewed publications, an extensive number of reports, and two joint Canada–United States Arctic operational capabilities: the Arctic Water Resources Vulnerability Index and the Arctic Adaptation Exchange Portal. She has also worked for two decades with remote and Indigenous communities to collectively pioneer unique, advanced, technology-enhanced capabilities that incorporate humans as an integral part of monitoring and observing systems for land, air, and maritime domains, honoring the diverse strengths of local, place-based, and traditional knowledge.

TOMAS DÍAZ DE LA RUBIA is the vice president for research and partnerships at the University of Oklahoma. Prior to this, he was the senior vice president for strategic initiatives and the chief scientific officer at Purdue University. Prior to joining Purdue, he was the innovation leader and a director in the energy and resources industry practice at Deloitte Consulting LLP. Prior to joining Deloitte in 2013, Dr. Díaz de la Rubia served as the chief research officer (2008-2012) and the deputy director for science and technology (2010-2012) at the Lawrence Livermore National Laboratory (LLNL), where he was the top executive responsible for the science and technology foundations of the laboratory's $1.6 billion program of national and nuclear security research. Reporting to the laboratory director, his role was to translate the director's executive vision and mission priorities into the research function of the laboratory. Dr. Díaz de la Rubia participated in many high-priority basic science, energy, defense, and other government technology programs and conducted business with the Department of Energy (DOE), DoD, DHS, and the Intelligence Community (IC). He is a consultant to the Defense Science Board. He holds both a B.S. (summa cum laude) and a Ph.D. in physics from the State University of New York at Albany.

VISHVA M. DIXIT is the vice president of early discovery research at Genentech. Additionally, he serves on the scientific advisory board of the Howard Hughes Medical Institute and the Gates Foundation. He has made many contributions to biomedicine and his early work on cell death is prominent in introductory textbooks of biology and medicine. He is an elected member of both the NAS and the NAM.

APPENDIX D

DONALD DUNCAN joined the Johns Hopkins University Applied Physics Laboratory in July 2004. He currently serves as the senior advisor in the Asymmetric Operations Sector. In this role, he focuses on development of cyber and information operations capabilities and systems. Dr. Duncan has an extensive background in systems engineering, development, and R&D management for telecommunications, and computer communications systems and networks. He retired from AT&T as Multi-Service Data Network Development vice president in 2002, after 29 years of service. From May 2002 until July 2004, he served as a fellow for the Technology Strategy for the IBM/Openwave Alliance. Dr. Duncan also served as a Reserve Officer in the U.S. Army Ordnance Corps, retiring with 28 years of service in 1997 with the rank of Lt. Col. Dr. Duncan has numerous publications on space-time physics and electron scattering. He is a member of the American Physical Society (APS), the Military Operations Research Society, and the Armed Forces Communications Electronics Association. He attended North Carolina State University and received a B.S. in physics. He then attended The University of Texas at Austin, where he received an M.A. and a Ph.D. in physics. He has also attended numerous U.S. Army Service Schools during his time as an Army Reserve Officer.

GERALD L. EPSTEIN joined the National Defense University's Center for the Study of Weapons of Mass Destruction in 2018 as a Distinguished Research Fellow. In that capacity he addresses challenges posed by nuclear, chemical, and biological weapons, particularly including the security implications of advanced life sciences, biotechnologies, and other emerging and converging technologies. Prior to arriving at National Defense University (NDU), he was the assistant director for Biosecurity and Emerging Technologies at the White House Office of Science and Technology Policy, where he served on detail from his position as the Deputy Assistant Secretary for Chemical, Biological, Radiological, and Nuclear Policy at DHS. Before returning to government service in 2012, Dr. Epstein directed the Center for Science, Technology, and Security Policy at the Association for the Advancement of Science (AAAS), which he joined in 2009. Dr. Epstein is a fellow of the AAAS and the APS and currently serves on the National Academies' Board on Life Sciences. He received an S.B. in physics and in electrical engineering from MIT and a Ph.D. in physics from UC Berkeley.

KATHLEEN FISHER is the chair of the Department of Computer Science at Tufts University. Previously, Dr. Fisher was a program manager at the Defense Advanced Research Projects Agency (DARPA) where she started and managed the HACMS and PPAML programs and a principal member of the technical staff at AT&T Labs Research. She received her Ph.D. in computer science from Stanford University. Dr. Fisher's research focuses on advancing the theory and practice of programming languages. She is a fellow of the Association of Computing Machinery (ACM) and a Hertz Foundation Fellow. Dr. Fisher is a council member of the Computing Community Consortium, a board member of the Computing Research Association (CRA), and the past chair of DARPA's ISAT Study Group. She is also the past chair of the ACM Special Interest Group in Programming Languages, the past co-chair of CRA's Committee on the Status of Women, a former editor of the *Journal of Functional Programming*, and a former associate editor of *TOPLAS*. She is a member of the board of trustees of Harvey Mudd College.

JAMES R. GOSLER is a senior fellow at the Johns Hopkins University Applied Physics Laboratory, where he provides strategic advice to the laboratory's senior leadership. His current and prior service also includes membership on various IC and DoD Boards, including the Defense Science Board. He is a former member of the National Academies' Naval Studies Board. Mr. Gosler's previous professional experiences include a 33-year career at Sandia National Laboratories. In 1996, he entered the Senior Intelligence Service at the Central Intelligence Agency (CIA) as the first director of the Clandestine Information Technology Office (CITO). This office was formed to establish Information Operations as a core and focused CIA discipline. Through the innovative integration of targeting, analysis, technology development, technical operations, and human operations, CITO's operational capability significantly augmented and complemented CIA's core operational competencies. Mr. Gosler earned a B.S. in physics and mathematics and an M.S. in mathematics.

LAURA M. HAAS is the dean of the College of Information and Computer Sciences at the University of Massachusetts Amherst (UMass). Her research explores the integration of data in the service of accelerating new discoveries from data. Prior to joining UMass, Dr. Haas spent 36 years at IBM, where she was rose to the level of IBM fellow. Within IBM, she served as the director of the Accelerated Discovery Lab (2011-2017); she was director of computer science at IBM's Almaden research center from 2005 to 2011, and had worldwide responsibility for IBM Research's exploratory science program from 2009 through 2013. From 2001 to 2005, she led the Information Integration Solutions architecture and development teams in IBM's Software Group. Dr. Haas received numerous IBM awards for her contributions. Before joining IBM, she studied applied mathematics and computer science at Harvard University and computer science at The University of Texas at Austin, where she received her Ph.D. in 1981. Dr. Haas is an ACM fellow and a fellow of the American Academy of Arts & Sciences, a member of the NAE, the chair of the National Academies' Computer Science and Telecommunications Board, and the past vice chair of the board of Computing Research Association.

ROBERT F. HALE is currently a senior fellow at the Center for Strategic and International Studies and a senior executive advisor at Booz Allen Hamilton, where he does international consulting on financial management issues. From 2009 until 2014, Mr. Hale served as the comptroller and the chief financial officer at DoD. During those years he managed $600 billion budgets in time of war, made significant improvements in defense financial management, and oversaw efforts by the department to minimize the problems caused by sequestration and a government shutdown. From 1994 to 2001, Mr. Hale served as the head of Air Force financial management. Mr. Hale also spent 12 years as the head of the defense group at the Congressional Budget Office. Early in his career he served as a Navy officer. Mr. Hale served as a commissioner on the recent National Commission on the Future of the Army and is a past member of the Defense Business Board. Since 2001, he has been a fellow in the National Academy of Public Administration where he has participated in several studies. He is Level 3 certified in Defense Financial Management and is a Certified Defense Financial Manager (with acquisition specialty). He has received numerous awards from DoD and the federal government for distinguished public service. Mr. Hale holds a B.S. from Stanford University in statistics, an M.S. from Stanford University in operations research, and an M.B.A. from The George Washington University.

DANIEL E. HASTINGS heads the Department of Aeronautics and Astronautics at MIT. Dr. Hastings, who earned a Ph.D. and an S.M. from MIT in aeronautics and astronautics in 1980 and 1978, respectively, received a B.A. in mathematics from Oxford University in England in 1976. He joined the MIT faculty as an assistant professor in 1985, advancing to associate professor in 1988 and full professor in 1993. As a professor of aeronautics and astronautics and engineering systems, Dr. Hastings has taught courses and seminars in plasma physics, rocket propulsion, advanced space power and propulsion systems, aerospace policy, technology and policy, and space systems engineering. Dr. Hastings served as the chief scientist to the U.S. Air Force from 1997 to 1999. In that role, he served as the chief scientific adviser to the chief of staff and the secretary and provided assessments on a wide range of scientific and technical issues affecting the Air Force mission. He was the chair of the Air Force Scientific Advisory Board from 2002 to 2005. He led several influential studies on where the Air Force should invest in space, global energy projection, and options for a science and technology workforce for the 21st century. He is a member of the NAE and a fellow of the American Institute of Aeronautics and Astronautics (AIAA), the International Astronautical Federation, and the International Council on Systems Engineering.

FRANCES S. LIGLER is the Ross Lampe Distinguished Professor of Biomedical Engineering in the Joint Department of Biomedical Engineering in the College of Engineering at North Carolina State University and the School of Medicine at the University of North Carolina at Chapel Hill. Until 2013, she was the senior scientist for Biosensors and Biomaterials at the U.S. Naval Research Laboratory in Washington, DC. Currently working in the fields of biosensors and regenerative medicine she has also performed research in biochemistry, immunology, microfluidics, and analytical chemistry. She has more than 400 full-length publications and patents, which have led to 11 commercial biosensor products. Elected an SPIE fellow in 2000, a fellow of AIMBE in 2011, and a fellow of AAAS in 2013, she also serves on the organizing committee for the World Biosensors Congress and the permanent steering committee for Europt(r)odes, the European Conference on Optical Sensors. Dr. Ligler has been recognized by the NAE with its Simon Ramo Founders Award. Dr. Ligler was elected to the NAE in 2005 "for the invention and development of portable optical biosensors, service to the nation and profession, and educating the next, more diverse generation of engineers." She is also a member of the National Academies' Committee on Science, Engineering, Medicine, and Public Policy. Dr. Ligler earned a B.S. in biology-chemistry from Furman University in 1972 and both a D.Phil. in biochemistry in 1977 and a D.Sc. in biosensor technology in 2000 from Oxford University. She was awarded honorary doctorates by the Agricultural University of Athens (Greece) in 2014 and Furman University (South Carolina) in 2018.

BERNARD S. MEYERSON is retired from IBM where he was an IBM fellow and served as IBM's chief innovation officer, driving technical strategy and corporate initiatives within IBM's Corporate Strategy Organization. In 1980, Dr. Meyerson joined IBM Research, leading the development of high-performance silicon:germanium communications technology. He founded and led IBM's highly successful Analog and Mixed Signal business, ultimately leading IBM's global semiconductor development. In 2006, he assumed leadership of strategic alliances for the Systems and Technology Group. Dr. Meyerson is a fellow of the APS, the Institute of Electrical and Electronics Engineers (IEEE), and a member of the NAE. His technical and business awards include the following: the Materials Research Society Medal, the Electrochemical Society Electronics Division Award, the IEEE Ernst Weber Award, the 2007 Lifetime Achievement Award from SEMI, and the 2011 Pake Prize of the APS (recognizing his combined original scientific research and subsequent business leadership). He holds a Ph.D. in physics from the City University of New York.

LISA J. PORTER[1] is the co-founder and the co-president of LogiQ, a company providing high-end management, scientific, and technical consulting services. She is also a member of the Riverside Research board of trustees, a not-for-profit organization chartered to advance scientific research for the benefit of the U.S. government and in the public interest. She was previously the Deputy Under Secretary of Defense for Research and Engineering, and in that role, she shared responsibility with the Under Secretary for the research, development, and prototyping activities across DoD. In prior roles, Dr. Porter served as the executive vice president of In-Q-Tel (IQT) and the director of IQT Labs, the president of Teledyne Scientific & Imaging, the first director of the Intelligence Advanced Research Projects Activity in the Office of the Director of National Intelligence, the associate administrator for the Aeronautics Research Mission Directorate at NASA, and as a program manager and senior scientist at DARPA. Dr. Porter holds a bachelor's degree in nuclear engineering from MIT and a doctorate in applied physics from Stanford University. She received the Office of the Secretary of Defense Medal for Exceptional Public Service, the NASA Outstanding Leadership Medal, the National Intelligence Distinguished Service Medal, the Presidential Meritorious Rank Award, and the DoD Medal for Distinguished Public Service.

PETER SCHIFFER is the Frederick W. Beinecke Professor of Applied Physics and a professor of physics at Yale University and a senior fellow at the Association of American Universities (AAU). At the AAU he works on major challenges facing the research enterprise in higher education, including the impact of the pandemic on research and efforts to address foreign government interference in university research. Dr. Schiffer earned his B.S. in physics from Yale University and Ph.D. in physics from Stanford University. He undertook postdoctoral work at AT&T Bell Laboratories before launching his faculty career at the University of Notre Dame. He later joined the faculty at The Pennsylvania State University, eventually serving as the associate vice president for research and the director of strategic initiatives. He subsequently served as a professor of physics and the vice chancellor for research at the University of Illinois at Urbana-Champaign, and then joined Yale University as the inaugural vice provost for research, the first university-wide senior research officer in the institution's history. Dr. Schiffer has also served in various leadership positions in scholarly organizations, and he has received multiple honors for his seminal research in emerging areas of the study of magnetism.

---

[1] Resigned from the committee on May 16, 2022.

ANTHONY J. VINCI is a managing director at a private equity and venture capital investment fund, an adjunct senior fellow with the Technology and National Security Program at the Center for a New American Security, a member of the board of trustees Technology Committee of MITRE, and a board member or advisor to multiple technology companies. He was a senior intelligence official and served as the chief technology officer and the associate director for capabilities at the National Geospatial-Intelligence Agency. Dr. Vinci has published and spoken extensively on the subjects of innovation, technology, and modernization in national security and the IC in *Foreign Affairs*, *The Atlantic*, and other publications. He received his Ph.D. in international relations from the London School of Economics and studied philosophy at Reed College and the University of Oxford. Dr. Vinci is a member of the Council on Foreign Relations and Business Executives for National Security.

MICHAEL S. WITHERELL is the director of Lawrence Berkeley National Laboratory and a professor of physics at UC Berkeley. Previously, he was the vice chancellor for research and held the Presidential Chair in Physics at the University of California, Santa Barbara (UCSB). Dr. Witherell served as the director of Fermi National Accelerator Laboratory (Fermilab), the largest particle physics laboratory in the country, from 1999 to 2005. From 1981 to 1999, he was a faculty member in the UCSB Physics Department. Dr. Witherell was elected to membership in the NAS in 1998 for his work in the application of new technologies that "profoundly influenced all subsequent experiments aimed at the study of heavy-quark states." In 2004, he received the U.S. Secretary of Energy's Gold Award, the highest honorary award of DOE. Dr. Witherell is a member of the NAS and a fellow of the APS, the AAAS, and the American Academy of Arts & Sciences. He currently sits on the National Academies' Committee on Science, Engineering and Public Policy. He received a Ph.D. from the University of Wisconsin–Madison in 1973 and B.S. from the University of Michigan, Ann Arbor, in 1968.